CAMBRIDGE MONOGRAPHS ON PHYSICS

GENERAL EDITORS

M.M. WOOLFSON, D.SC.
Professor of Theoretical Physics, University of York

J.M. ZIMAN, D.PHIL., F.R.S.
Professor of Theoretical Physics, University of Bristol

THE SCANNING ELECTRON MICROSCOPE

Top: Smith's Microscope, completed in 1958. *Bottom:* the Stereoscan manufactured by Cambridge Scientific Instruments Ltd.

Frontispiece

THE SCANNING ELECTRON MICROSCOPE

PART I

THE INSTRUMENT

C. W. OATLEY, F.R.S.

Emeritus Professor of Electrical Engineering
University of Cambridge

CAMBRIDGE
AT THE UNIVERSITY PRESS
1972

Published by the Syndics of the Cambridge University Press
Bentley House, 200 Euston Road, London NW1 2DB
American Branch: 32 East 57th Street, New York, N.Y. 10022

© Cambridge University Press 1972

Library of Congress Catalogue Card Number: 70–190413

ISBN: 0 521 08531 4

Printed in Great Britain
at the University Printing House, Cambridge
(Brooke Crutchley, University Printer)

CONTENTS

[v]

PREFACE

The modern scanning electron microscope is a relatively simple instrument to use and an operator with the minimum of training can obtain excellent micrographs of a wide range of specimens. Nevertheless, he has under his control a considerable number of variables. He must select an appropriate accelerating voltage for the electron beam, lens currents to provide the required demagnification and an aperature size to give a reasonable compromise between exposure time and resolution or depth of field. He can vary the angle at which the electron beam strikes the specimen and the distance of the latter from the final lens. The position and orientation of the collector can be changed and so, also, can the voltage applied between collector and specimen. Finally, he may wish to modify the instrument itself, to adapt it to a particular piece of research.

It is clear that, if he is to make the best possible use of his microscope, an operator should have a knowledge of the principles on which the design of the instrument is based and an understanding of the facts relating to the interaction of electrons with solid specimens. In the following pages I have attempted to provide this information in the simplest possible form. No prior knowledge of electron optics is assumed, nor of the theory of electron/solid interactions.

The scanning microscope was originally designed for the visual examination of specimens, using secondary or back-scattered electrons leaving the surface on which the incident electron probe impinged. It is with the basic instrument for this purpose, still the most important, that this book is concerned. No account is given of the scanning transmission microscope or of accessories such as that required for scanning electron diffraction. These matters, together with a number of special applications of scanning microscopy, will be dealt with in a second volume which my colleague Dr W. C. Nixon is at present preparing.

I take this opportunity of paying tribute to the successive generations of research students who have worked with me on the development of this fascinating instrument. Their contributions are acknowledge in the pages which follow, but special mention must be made of Dr D. McMullan and Dr K. C. A. Smith, whose work led to the construction of the first successful scanning microscope, and to Mr A. D. G. Stewart, who has played a major part in the commercial manufacture of the instrument. Finally, I wish to acknowledge my debt to the Directors of Cambridge Scientific Instruments Ltd, whose early faith in scanning microscopy contributed so greatly to subsequent progress.

January 1972   C. W. OATLEY
Cambridge

CHAPTER I

HISTORICAL INTRODUCTION

1.1. The principle of the instrument

The principle of the scanning electron microscope is illustrated schematically in fig. 1.1. A narrow beam of electrons which have been accelerated from a cathode C passes through electron lenses L_1, L_2, L_3 and is brought to a focus on the surface of the specimen S. The lenses are so placed that the diameter of the beam at S is very small and may be as little as 0.01 μm or less. A portion of the current

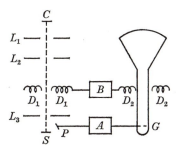

Fig. 1.1. The principle of the scanning electron microscope.

leaving S is collected by a plate P and conveyed to an amplifier A, the output of which determines the potential of the modulating electrode G of a cathode-ray tube. Thus the brightness of the spot on the face of this tube is controlled by the current reaching the collector P.

Current from a saw-tooth generator B passes through pairs of coils $D_1 D_1$ and $D_2 D_2$ respectively, causing the initial electron beam in the microscope and the spot on the face of the cathode-ray tube to be deflected. Two such systems are used to produce deflections at right angles to each other, so that the electron beam focused on the specimen S and the spot on the cathode-ray tube describe rectangular zig-zag rasters in synchronism. As the electron beam scans the surface of S, changes in composition, texture or topography at the point where the electrons strike the specimen cause variations in the electron current reaching the collector P and thus in the brightness of the cathode-ray tube spot. Thus there is built up on the screen a picture which is in some sense an image of S, though it is not immediately obvious that the picture will bear much resemblance to an optical image of the specimen surface. However, as we shall see later, this turns out to be the case, and it is on this fact that the usefulness of the scanning electron microscope largely depends.

By suitable choice of the numbers of turns on the various sets of deflecting coils, or by shunting those coils which deflect the electron beam in the microscope, it can be arranged that the size of the raster scanned on S is very much smaller than that on the surface of the cathode-ray tube. Then the final picture will be a magnified image of the object and, within limits, we can make the magnification as large as we please. In practice, as in most optical or electron-optical instruments, the useful magnification depends on the resolution that can be achieved but, with suitable objects, it may be as great as 100 000 times. On the other hand there are many objects for which a very much lower magnification is appropriate.

The above general scheme may be modified in various ways without changing the basic principle. Instead of collecting a fraction of the electrons which leave the specimen and travel to the collector P, an output signal may be obtained from variations in the total current flowing through S and returning to the cathode. Again, if the specimen is sufficiently thin, the collector may be placed on the side remote from the cathode and the output signal is then derived from variations in the current traversing the specimen. Finally, the component parts of the instrument, which are represented diagrammatically in fig. 1.1, may take a variety of forms. In the chapters which follow we shall discuss these modifications in detail.

1.2. The early history of scanning electron microscopy

The first instrument operating on the above principles was built by Knoll (1935) and further work was published by Knoll and Theile (1939). These investigators were primarily interested in studying secondary emission and the use of their apparatus as a microscope was incidental. In consequence they made no attempt to obtain an electron probe of very small cross-section and their electron beam was produced by an electrode structure similar to that commonly used in cathode-ray tubes. There were no demagnifying lenses and resolution was limited by the final beam diameter to something of the order of 100 μm.

The first true scanning electron microscope was built by von Ardenne (1938), though his apparatus differed in several respects from the scheme shown in fig. 1.1. In particular, no cathode-ray tube was used and the image was recorded photographically on a rotating drum. Furthermore, the instrument was designed for the study of very thin specimens by means of transmitted electrons.

The arrangement of von Ardenne's apparatus is shown diagrammatically in fig. 1.2. A narrow beam of electrons from the gun A passed successively through magnetic lenses B and C to form on the object D, a focused spot whose diameter was of the order of 0.01 μm. Electrons passing through the object fell on photographic film attached to a cylindrical drum E, which was caused to rotate and to progress axially, by means of a screw thread. The electron probe was deflected in two perpendicular directions by magnetic fields from coils F_1 and F_2 and the currents through these coils were controlled by potentiometers attached to the rotating drum. It was thus possible to arrange for the focused electron probe to describe a small zig-zag raster over the surface of the object, while the electrons passing through the object fell on a small area of the photographic film which effectively moved over the film in a corresponding raster some thousands of times as large.

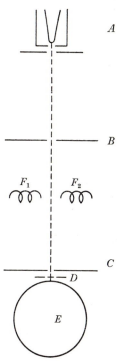

Fig. 1.2. The principle of von Ardenne's microscope.

Von Ardenne clearly realized the theoretical advantages and limitations of a system of this kind. In principle, the resolving power is ultimately limited by the aberrations of the electron lenses and therefore need be no worse than that in a conventional transmission electron microscope. In fact, as von Ardenne pointed out, chromatic aberration should be reduced in the scanning instrument because electrons which have penetrated the specimen, and which therefore have a range of velocities, are not subject to focusing by a further magnetic lens, as they are in a conventional microscope. However, it was also realized that the reduction of the final cross-section of the electron probe to a diameter of about 0.01 μm would necessarily limit the current in the probe to something of the order of 10^{-12} A and that, with such low current, the particulate nature of the electron probe would result in statistical fluctuations of the signal recorded on the photographic film. These fluctuations appear as background noise in the image and their amplitude can be reduced by increasing the time taken to record a picture, so that a larger number of electrons falls on each picture element. In von Ardenne's experiments the recording time was often as long as ten minutes or more.

Although von Ardenne's microscope was based on sound principles, it failed because its performance compared unfavourably with that of the conventional transmission instrument. The one advantage that could be claimed for the scanning microscope was that chromatic aberration might be lower when relatively thick specimens were being examined. Against this, the apparatus was much more complicated; no provision was made for direct observation of the image, so the instrument was much more difficult to adjust; the recording time was much greater; and, finally, the resolution actually obtained was not as good as that produced by a conventional electron microscope. Von Ardenne considered the possibility of using a scanning microscope in the manner depicted in fig. 1.1, in which secondary electrons from the front surface of a thick specimen are used to produce the image, but this line of investigation was not pursued.

Details of a new scanning electron microscope were published by Zworykin, Hillier and Snyder (1942) and these investigators were the first to demonstrate the possibilities of the instrument for the examination of the surface of an opaque specimen. At that time replica techniques were in their infancy, so competition from the conventional electron microscope was much less severe with opaque specimens than with thin specimens which could be examined directly by transmission.

The essential features of one form of this instrument are shown in fig. 1.3. Electrons from a tungsten filament F passed through a controlling grid G and were then accelerated by a potential difference of 10 000 V applied between F and the anode A. The electron beam so formed passed through electrostatic lenses L and M, which formed a greatly reduced image of the source, so that the spot focused on the object S had a diameter of the order of 0.01 μm.

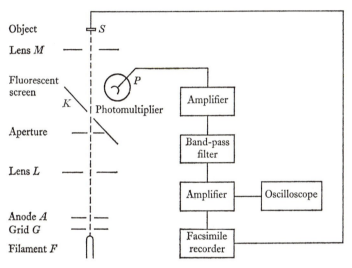

Fig. 1.3. Schematic diagram of the microscope of Zworykin, Hillier and Snyder.

S was maintained at a positive potential of about 800 V with respect to F, so that the incident beam struck the object with a velocity favourable to the production of secondaries. The secondary electrons were attracted back through lens M and, being of relatively low initial velocity, were brought to a focus shortly after leaving this lens. Thereafter they spread out to strike the surface of a fluorescent screen K in which there was a hole to permit the passage of the primary electron beam. Light from K fell on a photomultiplier P and the output from this device provided the signal from which the final image was built up. It will be shown later

(§3.4.6) that the combination of fluorescent screen and photo-multiplier adds very little noise to the signal while providing con-siderable amplification. Even so, noise resulting from the discrete nature of the electron stream falling on the object was very trouble-some and, as in von Ardenne's instrument, recording times of the order of ten minutes were needed. The final image was therefore recorded on a facsimile printer and, since this did not provide a convenient means of focusing the instrument, an oscilloscope was added to display the waveform of the output signal. The microscope was judged to be in focus when the highest frequency components in the waveform had their maximum amplitude: any defocusing caused an increase in the diameter of the incident electron probe and thus reduced the amplitudes of these components.

With such long recording times, amplification of the output signal would normally have required a d.c. amplifier and, at that time, the attendant problem of drift was by no means negligible. This difficulty was overcome by chopping the incident electron probe with a 3000 Hz square-wave voltage applied between grid and cathode, thus causing the output from the photomultiplier to become a modulated signal with a carrier frequency of 3000 Hz. Amplification of such a signal presented no difficulty since a.c. coupling could be used; moreover it was possible to include in the amplifier chain a band-pass filter of 700 Hz width, to exclude com-ponents of noise which lay outside the frequency range of the signal, as well as a non-linear amplifier with a rising amplitude characteristic to give greater contrast in the image. The signal was effectively demodulated within the facsimile recorder itself.

In the original instrument scanning was carried out mechanically, displacement of the object being produced by mechanical or hydraulic links controlled by the recorder. At a later stage greater precision was obtained by magnetic deflection of the incident electron beam, using saw-tooth circuits triggered by the facsimile recorder to produce the deflecting currents. To provide space for the deflecting coils two additional lenses were added to the original microscope column. With the final instrument an estimated resolu-tion of about 0.05 μm was obtained.

The scanning microscope built by Zworykin, Hillier and Snyder, which has just been described, represented a considerable advance in the art; moreover, it introduced a number of new techniques, some of which have proved valuable in more recent work. However, the instrument was not considered sufficiently promising to warrant further development and the project was discontinued. Many of the defects of this instrument can be attributed to the fact that the signal/noise ratio obtained with it was very poor by modern standards. While it is difficult to be dogmatic about the reasons for this, it seems likely that only a small fraction of the secondary electrons leaving the specimen actually contributed to the final signal and that the photomultiplier and amplifier introduced more noise than their modern counterparts would do. Whatever the reason, even micrographs obtained with long recording times were unacceptably noisy, while it was impossible to use short recording times of the order of one second and thus to provide a cathode-ray tube display which would enable the operator to select an area of the object for examination and to focus on it. A further probable defect, about which one cannot be certain, was that the method of collecting secondary electrons seems likely to have had an adverse effect on contrast and three-dimensional appearance in the image. Finally, the instrument was complicated and the recording system expensive.

We shall not describe in detail other early attempts to construct a satisfactory scanning microscope since they added nothing to the state of the art. Mention may, however, be made of a theoretical treatment published by Brachet (1946), in which he showed that, if the secondary electron current leaving the specimen could be collected and amplified without the introduction of additional noise, a resolution of 0.01 μm should be attainable. Davoine (1957) reported the construction of a special-purpose scanning microscope for the study of the secondary emission from stressed metals, but the resolution achieved was not better than about 2 μm.

1.3. The work of the Cambridge group; McMullan's microscope

In 1948 I initiated a programme of research on scanning electron microscopy in the Engineering Department of the University of Cambridge and I and my colleagues have continued to work in this

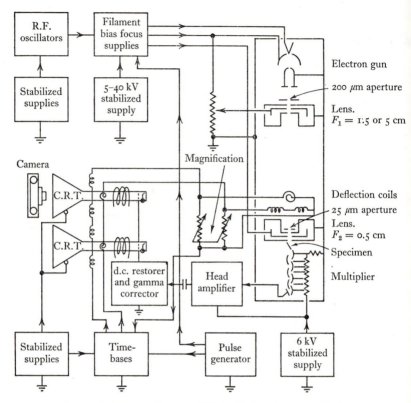

Fig. 1.4. Schematic diagram of McMullan's microscope. Redrawn by permission from McMullan (1953). *Proc. I.E.E.* B **100**, 245.

field since that time. The results of these researches will be considered in later chapters, but it is appropriate here to give some details of the first scanning microscope to be constructed at Cambridge. This instrument was designed by McMullan (1953) and the schematic diagram shown in fig. 1.4 is taken from a paper published by him.

The improvements introduced in McMullan's microscope were made possible by the achievement of a much better signal/noise ratio than had hitherto been obtained with opaque specimens. To ensure that a reasonable fraction of the electrons leaving the surface of the specimen should enter the collector, this surface was set at an angle of about 30° to the incident beam. The collector itself took the form of a secondary-emission electron multiplier which provided high amplification without the addition of much noise. Devices of this type had been known for many years, but the earlier multipliers had contained caesium-coated electrodes which could not conveniently be used in demountable vacuum systems. However, shortly after the war, multipliers with beryllium–copper dynodes became available (Baxter, 1949) and it was a detector of this kind that McMullan used. Although the above system has since been superseded, it represented a considerable advance on any previously used and has played a large part in the development of the scanning microscope. At the outset it was feared that the foreshortening of the final image brought about by allowing the incident electrons to strike the specimen obliquely, might be objectionable, but this has not proved to be the case and oblique incidence is commonly use in modern instruments.

With the improvement in signal/noise ratio, McMullan was able to reduce recording times and it became possible to obtain tolerable images with a frame period of one or two seconds. He therefore used a cathode-ray tube with a long-afterglow screen and was thus able to provide visual images which very greatly facilitated focusing and the selection of the appropriate area of the object. However, such tubes are not very satisfactory for photographic recording since they usually exhibit some degree of halation, which impairs definition. A second cathode-ray tube was therefore added for photographic purposes and sweep frequencies were arranged so that, with this tube, the recording time could be extended to several minutes to secure better integration of background noise.

Another refinement, suggested by McMullan and incorporated later, was a double deflection of the incident electron probe. The incident beam had to be scanned over the surface of the specimen and it was inconvenient to deflect the electrons after they had passed through the final lens, since very little room was available to accommodate the scanning coils at this point. It was therefore decided to mount the coils between the last two lenses and to use two pairs of coils for each coordinate direction to deflect the beam twice, as shown in fig. 1.5, so that it always passed symmetrically through the final aperture of the lens and thus kept aberrations to a minimum.

In earlier instruments it had been tacitly assumed that, with opaque specimens, contrast would result solely from variations in the secondary-emission coefficient of the surface. McMullan was the first to realize that other factors could have an important effect on contrast, but we shall defer consideration of these matters until

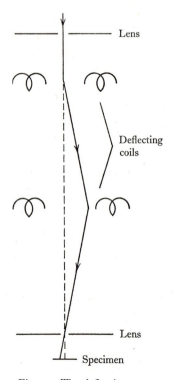

Fig. 1.5. The deflection system.

a later chapter. In his instrument the signal current was generated almost entirely by high-velocity back-scattered electrons, rather than by low-velocity secondaries.

1.4. Smith's microscope

Although McMullan's work was of the first importance in the development of the scanning electron microscope, his time at Cambridge was fully occupied with the design and construction of his instrument and he was not able to apply it to the examination of a wide range of specimens. This task was undertaken by K. C. A. Smith, who inherited McMullan's microscope and made numerous improvements to it (Smith, 1956). In particular, by altering the position of the collector and by keeping it at a positive potential with respect to the specimen, Smith was able to ensure the collection of slow secondary electrons, as well as the faster back-scattered ones. This improved the signal/noise ratio and gave better contrast in the final image.

By this time, the possibilities of the instrument were beginning to be appreciated by research workers in various laboratories and numerous specimens were examined for them. One in particular, a fibre of spruce, aroused great interest in the Pulp and Paper Research Institute of Canada and led that Institute to support the work at Cambridge and to request the construction of a microscope for its own use. The new instrument was built in the University Engineering Department, to Smith's design, and was shipped to Canada in 1958. It has given excellent service to the present day and has produced micrographs which are not greatly inferior to those obtained with the most modern instruments. (Smith, 1959 and Rezanowich, 1968.)

A photograph of Smith's Canadian microscope constructed in 1958 is shown in the frontispiece; this instrument is now in the Canadian National Museum of Science in Ottawa. From this photograph and from what has been said above, it will be appreciated that, by 1958, the scanning electron microscope had been established as a useful research tool. However, many of the factors affecting resolution and contrast were not understood at that time and subsequent research at Cambridge was largely concerned with matters such as these and with the application of the instrument to special problems. In 1965 this work culminated in the introduction by the Cambridge Instrument Company (now Cambridge Scientific Instruments Ltd) of a commercial model of the scanning electron microscope under the trade name Stereoscan. This is shown as the lower part of the frontispiece.

1.5. The transmission scanning microscope

It has already been mentioned that the scheme illustrated diagrammatically in fig. 1.1 does not constitute the only form of scanning electron microscope. If the object is sufficiently thin, the collector may be placed to collect electrons which have passed through the object, with or without scattering. The incident probe is scanned over the surface and the image is derived from the collected electrons in the usual manner.

For this arrangement to be possible, the thickness of the object must be limited to about a micron or less. Such an object can be examined in a conventional transmission microscope, so the question arises whether the transmission scanning microscope has any advantages to offer, to justify its greater complexity. During the past few years, this problem has been investigated in considerable detail by Crewe and his colleagues (Crewe, 1971), who have shown that, with certain objects, the scanning microscope can show detail which is not visible in the ordinary transmission instrument. Moreover, in principle, the resolution obtained with the former need not be inferior to that provided by the latter.

The great advantage offered by the scanning transmission microscope is the possibility of carrying out an energy analysis of electrons which have passed through each 'picture element' of the object and building up the final image from electrons whose loss of energy lies in any specified range. This facility arises from the fact that, in the scanning transmission microscope, the electron probe traverses the 'picture elements' of the object in sequence; in the ordinary electron microscope, the electron beam passes simultaneously through all elements of the object, and detailed energy analysis is then difficult. Since contrast in the final image may depend on the mechanisms of scattering and energy loss of the electrons, it is not surprising that the scanning microscope, with suitable processing of the transmitted electrons, can show detail which is invisible in the conventional instrument.

In the course of their work, Crewe and his collaborators have developed new types of electron gun and electron detector. In so far as these components have application to an ordinary scanning microscope constructed in accordance with the scheme of fig. 1.1, they will be described in the appropriate section of this book. We shall not, however, discuss the complete scanning transmission microscope, since the factors determining the optimum design of this instrument differ very materially from those which obtain when the specimen is opaque and electrons must be collected from the face on which the incident probe falls. The ordinary scanning microscope and the scanning transmission microscope are different instruments each with its sphere of usefulness. In this book we shall be concerned only with the former.

CHAPTER 2

FUNDAMENTAL LIMITATIONS ON DESIGN

2.1. Introduction

In this chapter we shall consider certain relations which set a limit to the performance which can be obtained with a scanning electron microscope. They can be expressed in the form of simple equations which enable us to estimate the best resolution that can be obtained when quantities such as the electron current density at the cathode, the aberrations of the electron lenses and the recording time are known. It does not follow that the calculated performance will be achieved in practice since, quite apart from incidental imperfections in the apparatus, we shall see later (§ 5.3.3) that resolution may be impaired as a result of penetration of incident electrons below the surface of the object. However, the basic equations serve as invaluable guides to the designer and enable him to assess the probable effect on the overall performance of the microscope of changes in the design of any particular component.

2.2. Current density in the incident electron probe

In any electron-optical instrument in which electrons from a cathode are brought to a focus in an image (in our case the electron probe) there is an upper limit to the current density which can be obtained in the image. This is given by the following equation, first derived by Langmuir (1937)

$$J = J_0[(eV/kT) + 1]\sin^2\alpha, \qquad (2.1)$$

where J and J_0 are the current densities in the image and at the cathode surface respectively, e is the electronic charge, k is Boltzmann's constant, V the potential difference between the cathode and the point where the image is formed, T the absolute temperature of the cathode and α the semi-angle of the cone of rays which converge to form a point of the image. The limitation arises from the spread of initial velocities of the electrons emitted by the cathode;

it is quite independent of the nature of the electron-optical system, though aberrations in this system may cause the current density in the image to fall short of the limit set by (2.1).

In any of the systems with which we shall be concerned α will always be very small and eV will always be much larger than kT. Hence the maximum current density in the image is given by

$$J = J_0 eV\alpha^2/kT. \tag{2.2}$$

We shall later consider how closely this limiting value can be approached in practice.

2.3. Fluctuations in the electron probe current

If we substitute representative values into (2.2), it becomes apparent that, with a tungsten thermionic cathode, the maximum electron current that can be focused into a probe of diameter 0.01 μm is of the order of 10^{-12} A. By way of illustration, let us assume that this probe scans over the surface of the object with a frame period of one second and that we are attempting to achieve a picture definition of 1000 lines so that information is required from 10^6 separate elements of the object surface. Under these conditions the number of electrons falling on each element of the object during a single scan is only about six. Moreover, electron emission from the cathode is a random process, so the number of electrons falling on a particular element will be subject to the usual statistical fluctuations and, with such a small average number of electrons, the fluctuation will be large. As a result there will be superimposed on the image produced by the scanning microscope a background of random noise and this proves to be one of the most important factors limiting the performance of the instrument.

The above simple discussion shows what steps might be taken to reduce the background noise, since the problem is essentially one of obtaining a larger number of incident electrons per picture element. Clearly, we should like to have a larger current density in the electron probe and (2.2) tells us that this can be achieved by increasing J_0 or by reducing T: we shall later enquire what can be done in these directions. We might also increase α, but this would lead to larger aberration in the final electron lens. For a given value of current density in the probe and a given probe diameter, we can increase the number of electrons per picture element only by reducing the number of lines in the picture or by increasing the time during which the picture is scanned. Obviously, these are conflicting requirements and, before a rational choice of values can be made, we need quantitative relations between the parameters involved. Very similar problems arise in the design of television camera tubes and we shall here follow a treatment given in that connection by Rose (1948).

With an electron gun of the type commonly used in microscopes, electrons from the cathode converge to form a 'crossover' of finite extent, where the current density reaches a maximum value. A detailed treatment of the electron gun and of the current density in the crossover will be given later (§3.1); here we assume the crossover to have a circular cross-section over which the current density is uniform, with zero current outside this area. Moreover, we take the current density within the crossover to have the limiting value given by (2.2). The crossover forms the object for subsequent demagnification by the electron lenses of the microscope and the final image so produced is the electron probe which scans the specimen under investigation. For the moment we ignore aberrations in the lenses so that, with the assumptions made above, the focused probe will also have circular cross-section over which the current density is uniform and is given by (2.2). Let d_0 be the diameter of this cross-section. With a given electron gun and hence a given area of the initial crossover, d_0 can be controlled by varying the strengths of the electron lenses.

Let the probe scan a square area of side D on the surface of the specimen and let the scan time be t. In general we cannot hope to resolve detail smaller than the cross-section of the probe, so we take $\frac{1}{4}\pi d_0^2$ to be the area of a picture element at the object. Then the number n of electrons falling on a picture element in a single scan is given by

$$n = \frac{J\pi d_0^2}{4e} \cdot \frac{d_0^2 t}{D^2}. \tag{2.3}$$

We shall shortly see that, to produce a satisfactory micrograph, n must be quite large. It will, however, be subject to statistical fluctuation because of the randomness of electron emission from the cathode, and the r.m.s. value of the fluctuation will be $\sqrt{n}$. We may therefore take the ratio

$$n/\sqrt{n} = \sqrt{n} \qquad (2.4)$$

as the basic signal/noise ratio for the element being scanned. Subsequent stages in the process of forming the final image cannot improve this ratio though they may make it worse. For the time being we suppose them to leave it unchanged.

The presence of background noise in the final image makes it difficult for the eye to distinguish areas of nearly equal brightness. Following Rose, we assert that, as a matter of experimental fact, the eye cannot distinguish an area of brightness B from an adjacent area of brightness $B \pm \Delta B$ unless the ratio of signal to noise is approximately five times the ratio of B to ΔB. We therefore write

$$\sqrt{n} \geqslant 5B/\Delta B, \qquad (2.5)$$

giving with (2.3)

$$25(B/\Delta B)^2 \leqslant J\pi d_0^4 t/4D^2 e, \qquad (2.6)$$

or substituting for J from (2.2)

$$25(B/\Delta B)^2 \leqslant J_0 \pi d_0^4 V t \alpha^2 / 4D^2 kT. \qquad (2.7)$$

2.4. The number of lines per picture

We have so far made no mention of the number N of lines per picture and to determine this quantity we work backwards from the display unit of the microscope. When looking at the final image, the eye will be unable to resolve distances much smaller than 0.01 cm so that, if the image is a square of side 10 cm, 1000 lines will give a picture of high quality. Under some circumstances a smaller number of lines may be tolerable while, with a larger display tube, a correspondingly larger number would be of advantage. In general, however, N is likely to lie between 300 and 2000.

Turning now to the object, it seems reasonable to arrange that the required number of lines is achieved without overlap and this condition is fulfilled by writing

$$N = D/d_0. \qquad (2.8)$$

However, this needs further discussion. Clearly, we must not make D/d_0 greater than N, since this would mean that some areas of the object were not receiving any primary electrons. On the other hand with a probe of circular cross-section, the condition of (2.8) does not ensure uniform distribution of the primary electrons over the scanned area; those portions of the object lying along the axis of each scan line will receive more electrons per unit area than portions lying midway between two adjacent lines. If the boundary of the probe really were sharply defined, some overlapping of adjacent lines would probably be desirable but, in practice conditions are rather different. In the first place, the crossover of the electron gun has a somewhat ill-defined boundary so that, even if there were no lens aberrations, its focused image on the specimen would have a diffuse edge. It is therefore necessary to consider rather more care-fully what is meant by the diameter d_0. We take it to be the diameter of that circular area in the focused probe over which, in the absence of lens aberrations, the current density would be roughly constant. This is admittedly rather imprecise but a more rigid definition would be of little value. We might, for example, define d_0 to be the diameter of that circular area within which eighty per cent of the total probe current lies; in practice, however, the variation of current density over the surface of the probe spot depends rather critically on the adjustment of the electron gun and is most unlikely to be known in any given case. An advantage of the definition that we have adopted for d_0 is that it is consistent with the use of this quantity in (2.2). We shall refer to d_0 as the diameter of the Gaussian probe spot.

We shall shortly see that lens aberrations are by no means negligible, so the effective diameter d of the focused probe spot is greater than d_0 and the fall-off of current density at its edge is still less sharp than it was in the crossover. Thus when we choose d_0 in accordance with (2.8), we are automatically allowing for appreciable overlap of successive lines scanned on the object and this equation therefore represents a reasonable approximation to the ideal of uniform distribution of the primary electrons over the scanned area. In view of other uncertainties to be mentioned below, any ambiguity in the exact meaning of d_0 or lack of knowledge of the precise degree of overlapping of successive scan lines is of minor importance.

We may now combine (2.7) and (2.8) to obtain an expression for the smallest value of the Gaussian spot diameter which meets our conditions. We find

$$d_0^2 = \frac{100(B/\Delta B)^2 \, kTN^2}{J_0 \, \pi Vt\alpha^2}.$$

(2.9)

We recall that this lower limit to the value of d_0 is set solely by the necessity to provide a sufficient number of incident electrons per picture element to enable us to discriminate in the final image between an area of brightness B and an adjacent one of brightness $B \pm \Delta B$, in the presence of background noise caused by statistical fluctuation of the electron emission from the cathode. The equation takes no account of any happenings after the arrival of the incident electrons at the surface of the specimen and some of these must now be considered.

As a rule the incident electrons in a scanning microscope have energies of some thousands of electronvolts and the corresponding secondary-emission coefficient of the specimen will often be less than unity. In such cases, the fluctuation noise will be determined by the number of electrons per picture element which produce secondaries in a single scan, rather than by the total number which strike the element, since incident electrons which produce no secondaries contribute nothing to the output signal. Thus the signal/noise ratio will be smaller than the value we have calculated hitherto and, to compensate for this, d_0 must be made larger than the value given by (2.9). Even if the average secondary-emission coefficient is appreciably greater than unity, some of the incident electrons will produce no secondaries, while others produce more than one. Statistically, the fluctuation noise is determined primarily by the number of discrete pulses of electron current recorded from each picture element, and the fact that some of the pulses have been increased in amplitude (because one primary produces several secondaries) in no way compensates for the overall reduction in the number of discrete pulses. Again the signal/noise ratio has been reduced and d_0 must be made larger.

Two further effects act in the same direction. The collector will normally pick up only a fraction of the electrons leaving the specimen, so the number of discrete pulses recorded from each picture element will be reduced still further. Clearly, the magnitude of the reduction depends very much on the size and position of the collector and on the field applied between specimen and collector; under certain circumstances virtually all of the electrons leaving the specimen contribute to the output signal. Finally, whatever type of amplifier follows the collector is likely to add noise from itself. With the best types of amplifier the added noise is almost insignificant while with some other arrangements, which are useful in particular cases, the added noise may be appreciable in comparison with that resulting from the randomness of the cathode emission.

At a later stage we shall be discussing some of these factors in greater detail, but it is clear that their overall effect will depend very greatly on the particular experimental arrangement that is being used. However, it seems unlikely that, even in the most favourable circumstances, the reduction in the number of discrete pulses per picture element will be less than a factor of about three. It is therefore convenient to re-write (2.9) empirically as

$$d_0^2 = \frac{100\beta(B/\Delta B)^2 kTN^2}{J_0 Vt\alpha^2}, \qquad (2.10)$$

where the factor π has been dropped from the denominator and a constant β added in the numerator. β is unlikely to be less than unity and may be very much larger. Once again we note that (2.10) results entirely from a consideration of fluctuation noise and takes no account of lens aberrations.

2.5. Lens aberrations

Because the electron lenses are imperfect, a number of aberrations are introduced and, as a result of these, each point of the focused Gaussian image is expanded into a circle of least confusion in an adjacent plane. As stated previously, the net result is to make the effective diameter d of the probe larger than the diameter d_0 of the Gaussian image, and to produce greater variation of current density over the effective cross-section of the probe. We should expect this to impair resolution since d, rather than d_0, should now be taken as the diameter of a picture element at the surface of the specimen. However, (2.10) should still be valid, since the number of electrons which, with perfect lenses, would have passed through a probe cross-section of diameter d_0, now pass through the actual cross-section of diameter d. The area of a picture element has increased, but the number of electrons per picture element remains the same.

We first consider the diameters of the circles of least confusion into which each point of the Gaussian image would be expanded if the various aberrations were acting separately. Quoting standard results from electron optical theory, which we shall later discuss in greater detail, we have:

(a) As a result of spherical aberration alone (§ 3.2.3), the diameter of the circle of least confusion is

$$d_s = \tfrac{1}{2}C_s\alpha^3, \qquad (2.11)$$

where C_s is the spherical aberration coefficient of the final lens and α, as before, is the semi-angle of the cone of rays converging to form the Gaussian image point.

(b) Similarly, for chromatic aberration alone (§ 3.2.4), the diameter of the circle of least confusion is

$$d_c = C_c\alpha\delta V/V, \qquad (2.12)$$

where C_c is the chromatic aberration coefficient of the final lens, δV electronvolts the spread in energy of the electrons entering the lens and V the mean energy of these electrons. δV may arise partly from instability of the power supply but the Maxwellian distribution of initial velocities of electrons emitted by the cathode is usually a more important factor.

(c) Since electrons have an effective wavelength λ, diffraction effects will occur and the diameter of the central disc of the diffraction pattern (see, for example, Ditchburn, 1952), is given by

$$d_f = 1.22\lambda/\alpha. \qquad (2.13)$$

Substituting for λ in terms of the electron energy V in electronvolts,

$$d_f = 1.22 \times 10^{-10}\alpha^{-1}\sqrt{(150/V)}. \qquad (2.14)$$

(d) There will be aberrations caused by lenses other than the final one. However the images of the crossover produced by these earlier lenses have diameters so much larger than d_0 that the aberrations are negligible in comparison.

(e) There may be astigmatism resulting from eccentricity or misalignment of the lenses. In a well constructed microscope astigmatism should be almost negligible; if present, it can be corrected by a stigmator, so we shall not include it here.

In quoting the above results we have referred to circles of least confusion. It will be appreciated, however, that the area over which electrons from a Gaussian image point are spread as a result of a particular aberration is not, in fact, a circular disc with sharp boundary and uniform current density. Thus there is some uncertainty as to the exact effective values to be assigned to d_s, d_c and d_f, though the equations should reliably indicate the relative importance of the different aberrations. In practice the various aberrations are present simultaneously and any detailed assessment of their combined effect would be a matter of very considerable complexity. In view of the uncertainties previously mentioned, it is sufficiently accurate for our purpose to write for the effective diameter of the probe

$$d^2 = d_0^2 + d_s^2 + d_c^2 + d_f^2. \tag{2.15}$$

This method of combining the aberrations has often been used and is in reasonable agreement with experimental results.

2.6. The optimum aperture and the limit of resolution

Substituting from (2.10), (2.11), (2.12) and (2.14), equation (2.15) can be written in the form

$$d^2 = P\alpha^{-2} + Q\alpha^2 + R\alpha^6, \tag{2.16}$$

where
$$P = \frac{2.33 \times 10^{-18}}{V} + 100 \left(\frac{B}{\Delta B}\right)^2 \cdot \frac{\beta k T N^2}{J_0 V t}, \tag{2.17}$$

$$Q = (C_c \delta V/V)^2, \tag{2.18}$$

$$R = (\tfrac{1}{2}C_s)^2. \tag{2.19}$$

Differentiating (2.16) to find the value of α which makes d a minimum, we find
$$\alpha_{opt}^4 = [\sqrt{(Q^2 + 12PR)} - Q]/6R, \tag{2.20}$$

and the value of d obtained when this value of α is used is taken to be the smallest distance that the microscope is likely to be able to resolve.

Attention has already been drawn to various uncertainties that arise in the application of the above equations and there are others which have not been mentioned. For example, we shall see later that resolution is affected by the nature of the specimen under examination and, in particular, by the penetration of incident electrons into this specimen. Then again we are unlikely to know the exact value of β in (2.10) and the value that we assign to the ratio $B/\Delta B$ is to some extent arbitrary; it is likely to depend on the detailed structure present in the specimen and on the quality of the micrograph required. In this connection, it should be recalled that the criterion adopted for the minimum number of electrons per picture point at the object (2.5) merely ensures that an area of brightness B can be distinguished from an adjacent area of brightness $B \pm \Delta B$; it certainly does not guarantee a noise-free micrograph.

In view of these doubts it is pertinent to enquire how far the equations that we have developed provide a reliable guide to the resolution that will be obtained in practice. Before answering this question it is instructive to insert some typical numerical values into our equations to get an idea of the relative magnitudes of the different effects that we have considered. We shall see that some of these effects can be neglected in almost all cases and that the equations can then be rewritten in simpler forms.

2.7. Approximations that are valid when $d > 0.01\ \mu\mathrm{m}$

Consider a microscope operating under conditions specified by the following values. Though not necessarily the best that can be achieved, these represent typical modern practice.

$$C_c = 8 \times 10^{-3}\,\mathrm{m}, \quad C_s = 2 \times 10^{-2}\,\mathrm{m}, \quad V = 2 \times 10^4\ \text{volts},$$

$$T = 2800\,^\circ\mathrm{K}, \quad N = 1000, \quad J_0 = 2 \times 10^4\,\mathrm{A/m^2},$$

$$\delta V = 1\ \text{volt}, \quad B/\Delta B = 10, \quad \beta = 10, \quad t = 1000\ \mathrm{s}.$$

Substituting these values in (2.17), (2.18), (2.19), and (2.20), we obtain

$$P = (1.12 \times 10^{-22}) + (0.97 \times 10^{-20}), \tag{2.21}$$

$$Q = 1.6 \times 10^{-13}, \tag{2.22}$$

$$R = 10^{-4}, \tag{2.23}$$

$$\alpha_{\mathrm{opt}}^4 = [\sqrt{(Q^2 + 12PR)} - Q]/6R$$

$$= \frac{\sqrt{[(2.56 \times 10^{-26}) + (1.16 \times 10^{-23})]} - 1.6 \times 10^{-13}}{6 \times 10^{-4}}$$

$$= 0.54 \times 10^{-8}, \tag{2.24}$$

or $\quad \alpha_{\mathrm{opt}} = 0.86 \times 10^{-2}$ radian. $\tag{2.25}$

Examination of (2.21) shows that the first of the two terms, of which P is the sum, is negligible compared with the second. This first term arises from diffraction effects, which therefore have little or no influence on the optimum aperture. Similarly from (2.24) it appears that the terms in Q, which arise from chromatic aberration, are relatively unimportant in the determination of α_{opt}.

Turning now to (2.16) and substituting from (2.25), we find

$$d^2 = P\alpha^{-2} + Q\alpha^2 + R\alpha^6,$$
$$= (1.31 \times 10^{-16}) + (1.18 \times 10^{-17}) + (0.41 \times 10^{-16}), \quad (2.26)$$

whence
$$d = 1.35 \times 10^{-2}\,\mu\text{m}. \quad (2.27)$$

Once again we see that the middle term in (2.26), resulting from chromatic aberration, is a good deal less important than the other two in determining the final probe diameter. We therefore, conclude that, under the conditions of operation postulated above, both diffraction and chromatic aberration may be neglected without serious effect on the overall result. It is to be noted also that the conditions postulated lead to a minimum resolved distance of $1.35 \times 10^{-2}\,\mu\text{m}$, in good agreement with experimental results under the assumed conditions.

Suppose now that the conditions of operation of the microscope considered above are varied, while the aberration constants of the lenses remain the same. In an effort to secure the best possible resolution we might try to reduce the value of the second term on the right-hand side of (2.17). This might be done by increasing t; by accepting a smaller image and hence a smaller value of N; by arranging a more efficient collection of secondary electrons, to reduce β; or by accepting a smaller value of the ratio $B/\Delta B$. Changes in these directions must eventually lead to a situation where the second term in the equation for P, (2.17), is no longer negligible compared with the first and diffraction must be taken into account. Similarly, as P decreases, so also will α_{opt} and, since chromatic aberration is proportional to α while spherical aberration varies as α^3, the former will eventually become more important than the latter. However, we shall see later that, with opaque specimens and with values of d appreciably less than 0.01 μm, resolution may well be limited by effects arising from penetration of the incident electrons into the specimen rather than by the diameter of the probe.

Thus our approximations are unlikely to introduce serious error so long as purely electron-optical considerations are dominant.

In the foregoing paragraph we have investigated the effect of reducing P. A condition much more likely to occur is that in which the microscope is used with a magnification below its maximum useful value and with correspondingly reduced resolution. This will result in larger values of P, α_{opt} and d, and both diffraction and chromatic aberration will be quite negligible. Still another condition that is worthy of consideration is that in which the microscope is operated with an unusually low value of the accelerating voltage for the incident electrons. This will not affect the relative magnitudes of the two terms on the right-hand side of (2.17) since V occurs in the denominator of each. Thus diffraction can still be neglected. However, reduction of V does cause an increase of chromatic aberration, which depends on the ratio $\delta V/V$, and at some point this aberration will no longer be negligible. For a microscope operating under the conditions stated at the beginning of this section, except that V is 2kV instead of 20kV, neglect of chromatic aberration leads to a value of α_{opt} of 11.6 instead of 8.6 milliradian, while d becomes $3.1 \times 10^{-2}\,\mu$m instead of $5.0 \times 10^{-2}\,\mu$m.

Summarizing the above discussion we may say that, with an opaque object, diffraction and chromatic aberration may be neglected without serious error under almost all conditions under which a scanning microscope is likely to be operated. A possible exception occurs when the instrument is operated with a very low accelerating voltage.

The above statement is *not* true when a scanning microscope is used with a transparent specimen and when the output signal is derived from electrons which pass through the specimen.

2.8. The approximate equations

For the great majority of cases in which diffraction and chromatic aberration can be neglected, (2.16), (2.17), (2.19) and (2.20) can be rewritten as

$$d^2 = P\alpha^{-2} + R\alpha^6, \tag{2.28}$$

where

$$P = 100\left(\frac{B}{\Delta B}\right)^2 \frac{\beta kTN^2}{J_0 Vt}, \tag{2.29}$$

$$R = (\tfrac{1}{2}C_s)^2, \tag{2.30}$$

and
$$\alpha_{opt}^4 = \sqrt{(12PR)/6R} = \sqrt{(P/3R)}, \qquad (2.31)$$

whence
$$\alpha_{opt} = (P/3R)^{\frac{1}{8}} = 1.04 P^{\frac{1}{8}} C_s^{-\frac{1}{4}}. \qquad (2.32)$$

Substituting this value of α_{opt} in (2.28), we obtain

$$d = 1.3 P^{\frac{3}{8}} R^{\frac{1}{8}}$$

or
$$d = 1.1 P^{\frac{3}{8}} C_s^{\frac{1}{4}}. \qquad (2.33)$$

It is convenient to derive an expression for the total current I in the incident probe. From (2.2)

$$I = \pi d_0^2 J/4 = \pi d_0^2 J_0 V e \alpha^2 / 4kT, \qquad (2.34)$$

which by virtue of (2.10) and (2.29), becomes

$$I = 100\pi (B/\Delta B)^2 \beta N^2 e/4t$$
$$= \pi P J_0 V e/4kT. \qquad (2.35)$$

Eliminating P from (2.33) and (2.35), we get

$$I = 0.6 d^{\frac{8}{3}} J_0 V e / kT C_s^{\frac{2}{3}}$$
$$= 7 \times 10^3 d^{\frac{8}{3}} J_0 V / C_s^{\frac{2}{3}} T. \qquad (2.36)$$

In a similar manner, by eliminating P from (2.29) and (2.33), we obtain the following useful expression for the time of scan, t:

$$t = 1.8 \times 10^{-21} (B/\Delta B)^2 \beta T N^2 C_s^{\frac{2}{3}} / J_0 V d^{\frac{8}{3}}. \qquad (2.37)$$

The approximate forms of the equations that have just been derived show much more clearly than the earlier exact forms the relations between the quantities involved. We, therefore, return to the question raised in the preceding section as to the extent to which the equations are invalidated by uncertainties in β and in the ratio $(B/\Delta B)$. First we note that the primary effect of these uncertainties is to cause a corresponding uncertainty in the value of P. However, P occurs to the one-eighth power in (2.32), so a ten-fold error in P causes an error of only about thirty-five per cent in α_{opt}. Thus we may use this equation with considerable confidence. To some extent the same is true of (2.33), though here a ten-fold error in P will alter d by a factor of between two and three. Clearly we must not place too much reliance on the absolute value of d given by this equation. In (2.36) and (2.37) the uncertainties in I and t respectively will be the same as that in P, so these equations can be expected to yield approximate values only.

The real importance of these equations, however, lies not so much in their use to obtain absolute values of α_{opt}, d, I and t, as in the fact that they enable us to predict the variations in these quantities when other parameters are changed. For this purpose, any uncertainty in P is irrelevant so long as a change in the parameter under consideration does not involve a change in the uncertainty.

We end this chapter by re-stating the conditions under which (2.28) to (2.37) may be expected to apply. Thus:

(a) They represent limits to the operation of a scanning electron microscope which are set by lens aberrations and by fluctuation noise in the incident electron probe. They take no account of effects resulting from penetration of the incident electrons below the surface of the specimen, from electrical interference or from vibration. Any or all of these may cause the performance of the microscope to be worse than that predicted by the equations.

(b) The approximate forms of the equations should normally be valid so long as d is not less than about 0.01 μm.

When a scanning electron microscope is used to examine a transparent specimen and a signal is derived from electrons which have passed through the specimen, values of d much less than 0.01 μm can be used and the equations would then be seriously in error.

(c) The equations are derived on the assumption that the accelerating voltage in the microscope is of the order of 20 kV and that the final lens of the instrument is a magnetic lens of good quality with reasonably low aberration coefficients.

Any factors causing the aberrations to increase would make the approximations less valid, though the errors introduced would be unlikely to be very serious. Such a factor might be the use of a final magnetic lens of long focal length to give a long working distance; or the use of a particularly low accelerating voltage.

CHAPTER 3

THE COMPONENT PARTS OF THE SCANNING MICROSCOPE

In this chapter we shall consider the design of each of the major component parts of a scanning electron microscope. Although present practice will be clearly indicated, the principles of design will be discussed from a rather broader point of view, in the hope that the treatment will cover future developments in the state of the art. The scanning electron microscope is still a relatively new instrument and further uses for it are continually being found. There is every reason to expect that, in the years to come, special microscopes will be built for particular purposes and, in such instruments, there may well be a use for techniques which would be out of place in a general-purpose microscope. Thus, in dealing with components, it seems desirable to separate basic principles from techniques which, at the present time, are convenient.

Throughout the chapter a scanning microscope is taken to be an instrument in which the output signal is derived from electrons leaving the specimen by the same surface as that through which the primary electrons entered. The transmission scanning microscope is considered to be a separate instrument which is outside the present discussion.

3.1. The electron gun

3.1.1. The triode gun with thermionic cathode

A type of electron gun commonly used in electron-optical instruments is shown schematically in fig. 3.1. Thermionic electrons from a plane cathode C are accelerated by the field produced by the anode A, which is maintained at a high positive potential, usually greater than 1 kV, with respect to C. Between C and A is a third electrode G, which is held at a negative potential with respect to C and which we shall term the grid; it is also known as the modulator or the Wehnelt electrode. Both G and A consist of metal plates with circular holes through which the electron beam passes and these holes may or may not be of the same size. The whole system is symmetrical about an axis perpendicular to OC.

A complete analysis of such a system presents formidable difficulties, but its general mode of operation is as follows. The electric field produced by the three electrodes has symmetry about the axis OP, so the electrodes form an electron lens and, for the moment, we shall neglect any aberrations that it may have. Then an image of the cathode is formed in some plane PB.

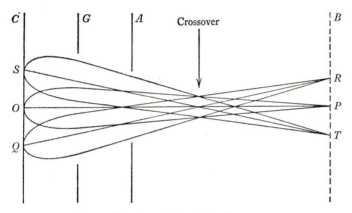

Fig. 3.1. The thermionic triode electron-gun.

Consider first those electrons which leave the cathode at the point O on the axis of the system. If these electrons had zero initial velocities, they would all move off in the direction of the applied field and would travel directly from O to the image point P. In actual fact they are emitted with a Maxwellian distribution of velocities and so move away from O in a variety of directions. Not many of the electrons will have large initial radial velocities, so the majority of those leaving O will have trajectories contained within a bundle of the form shown, and will be brought to a focus at P. Similarly, electrons leaving the cathode at points Q and S will be focused to image points R and T respectively. We now turn our attention to the overlapping of the three bundles of trajectories drawn in the figure and we see that they all pass through a common area, known as the crossover, which forms the minimum cross-section of the electron beam considered as a whole. Moreover, from each point of the crossover, the trajectories of electrons moving to the right form diverging pencils, so the crossover itself may be regarded as a small electron source. In the scanning microscope it is the demagnified image of such a crossover which forms the electron probe impinging on the specimen.

It must be understood that the above picture of the operation of a triode gun represents a highly idealized situation. It assumes that the crossover is formed in a field-free region where the electrons are travelling in straight lines. Under some conditions, however, the crossover may occur in the region between anode and cathode where trajectories are curved; the electrons will not be travelling in straight lines until they have passed beyond the anode. In this case we produce back the straight portions of the trajectories to a virtual crossover, from which each pencil of electron rays appears to have originated. This virtual crossover may sometimes lie behind the cathode. A complete analysis of the electron gun presents great difficulties and the problem has not yet been completely solved. The lens formed by the electrodes is likely to have quite large aberrations, the electron trajectories will be affected by initial velocities and, in many cases, by space charge. These difficulties need not concern us. When we speak of a crossover, we imply no more than that, as seen from the field-free region beyond the anode, there exists an effective source from which the electron pencils appear to diverge in straight lines. This source is not the cathode and it may be real or virtual: its boundary is unlikely to be clearly defined and its effective size and position must be determined by experiment.

If the electrons had been emitted with zero initial velocities, each of the trajectory bundles in fig. 3.1 would have degenerated into a straight line and the crossover would have had infinitely small area. The current density in the crossover would then have been infinitely great, thus violating Langmuir's law (§2.1). The foregoing considerations provide us with a simple physical picture of the way in which the current density in an electron-optical system is limited by initial velocities and why the limit depends on the ratio Ve/kT. The r.m.s. value of the energy resulting from the radial components of initial velocity is kT, while the forward energy contributed by the applied field is Ve. As the latter increases, the effect of the former on the size of the crossover (and hence on the current density) will be reduced. For detailed proofs of Langmuir's formula, from different points of view, reference may be made to Langmuir (1937), Pierce (1949) and Moss (1968).

Since the radial components of the initial electron velocities have a Maxwellian distribution, we might surmise that the radial distribution of current density in the crossover would be Gaussian. In practice, lens aberrations are rarely negligible but the distribution of current density usually approximates quite closely to the Gaussian law.

So far, we have accepted without question that the electrons emitted from the cathode have a Maxwellian distribution of initial velocities corresponding to the absolute temperature T of the cathode but, in reality, the matter is more complicated. According to the Richardson–Dushman equation, the saturated thermionic current density from a pure, uniform, plane metal surface at temperature T is given by

$$J_{sat} = AT^2(1-r)\exp(-e\phi/kT), \tag{3.1}$$

where A is a constant, ϕ the work function of the metal, e the electronic charge, and k Boltzmann's constant. The factor $(1-r)$ is included to allow for the possibility that a fraction r of the electrons in the metal, which approach the surface with sufficient normal velocities to enable them to surmount the surface potential barrier, are nevertheless reflected. In the present state of our knowledge it is not possible to calculate r; there are reasons for believing it to be small and it is often neglected. If r is a constant, it can be shown that both the normal and the tangential components of the initial velocities of the emitted electrons will have Maxwellian distributions, each with mean energy kT, so that the total initial mean energy is $2kT$.

Experiments by Nottingham (1939) indicate that the distribution of velocities of emission is not truly Maxwellian and that there is a deficiency of slow electrons. He has suggested that this might be explained if r were strongly dependent on normal velocity, though there is no theoretical reason to expect this. Nottingham's results have been confirmed by Hutson (1955) who experimented with a single-crystal tungsten emitter, but their work has been criticized by Herring and Nichols (1949) and by Smith (1955). When we turn to composite emitters such as oxide-coated cathodes, further complications arise. The surface of an emitter of this kind is seldom uniform and most of the electrons come from islands of low work function. Even a tungsten filament is usually polycrystalline and the different crystal faces have different work functions. There may also be effects resulting from the voltage drop caused by the passage of current from the core to the surface of the coating (Hadley, 1953). Finally, Moss (1961) has shown that, when the surface of a cathode is rough and the emission is temperature-limited, the external accelerating field may penetrate cavities in the surface and cause abnormally high apparent tangential emission velocities.

None of these complications seriously affects the picture given earlier of the operation of the electron gun. They may modify the distribution of current in the crossover, though not usually to an important extent and they may cause a change in the size of the crossover which, in any case, has to be determined by experiment. They do, however, lead to some uncertainty in the effective temperature T to be used in Langmuir's equation and we must bear this in mind when considering the efficiencies of different types of electron gun.

So far nothing has been said about the effect of the negative voltage V_g applied to the grid G of the electron gun in fig. 3.1. This voltage reduces the accelerating field at the surface of the cathode and has greatest effect on those areas of the cathode which are furthest from O. Over areas sufficiently remote from O the total field is retarding and no electron current is drawn from these areas. As V_g is made increasingly negative with respect to the cathode, the area from which electrons can escape gets smaller and smaller until it degenerates to a point at O. This occurs when V_g has the negative cut-off value V_c and the cathode current I_c is then zero. The general relation between I_c and V_g is of the form shown in fig. 3.2.

3.1.2. The complete electron-optical system of the microscope

Before dealing with particular types of electron gun it is convenient to give some consideration to the complete electron-optical system of which the gun forms a part, so that we can decide what characteristics are desirable in the gun. The whole system is shown

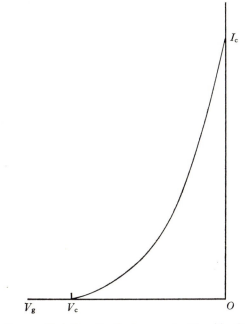

Fig. 3.2. Variation of cathode current with grid voltage.

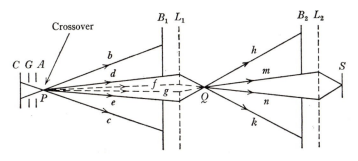

Fig. 3.3. The electron-optical system of the microscope.

schematically in fig. 3.3. The triode gun, with electrodes C, G and A, forms a crossover at P and this serves as an object for the first demagnifying lens L_1, which produces a reduced image of the crossover at Q. Q is the object for the second lens L_2, which produces a still further reduced image of the crossover on the surface of the specimen at S. Since the distribution of the current density at the crossover is roughly Gaussian, there will be no sharp boundary to

3

the cone of electron paths leaving the crossover, but we may take the paths b and c to represent the cone within which, say, ninety per cent of the current is included. An aperture B_1 rejects most of this current, so that the image at Q is formed from electron paths lying between d and e. Similarly, a second aperture B_2 restricts the electrons passing through L_2 to those paths lying between m and n. Producing m and n backwards along the dotted lines shown, it is apparent that, of the electrons leaving P, only those whose paths lie within the cone bounded by f and g actually contribute to the final image at S. It is also clear that this cone is defined by B_2 and that, for this purpose, the aperture at B_1 is superfluous. It is inserted to hold back a high proportion of unwanted electrons which might cause electrical leakage or other troubles if allowed to penetrate too far along the microscope column. Such an aperture is known as a spray aperture; it plays no part in the electron optics of the system.

An actual system may differ in a number of ways from the schematic diagram of fig. 3.3. The apertures will not necessarily be placed in the positions shown, relatively to L_1 and L_2, and the defining aperture at the final lens will be much smaller than any spray apertures that may be used. The lenses themselves may be thick lenses rather than the thin ones shown and there may be three lenses instead of two. In any case, distances and focal lengths will be such that the demagnification produced by each lens is very much greater than that shown in the diagram. It follows that the cone of paths included between f and g will be of very much smaller angle than the diagram suggests. Since only those electrons whose paths lie within this cone are of any value for our purpose, the first requisite of the electron gun is that it should concentrate as much current as possible into this extremely narrow pencil. Such a gun is known as a narrow-angle gun to distinguish it from the quite different type used in microwave tubes such as klystrons.

We have already seen in chapter 2 that the size of the electron probe at S is determined partly by lens aberrations and partly by the requirement of adequate current in the probe; and we have discussed the way in which these two limitations interact to determine the optimum size of the aperture at the final lens. In this discussion it was assumed that the current density at the surface of the specimen might, in the limit, reach the value given by Langmuir's equation. We now see that the realization of this hope depends on the efficiency of the gun and we need some means of relating conditions at the specimen to conditions at the crossover. For this purpose we ignore aberrations in the demagnifying lenses since these are independent of the properties of the gun and are, in any case, allowed for separately.

We wish to make use of a theorem in light optics, often called the Helmholtz–Lagrange theorem, which has its counterpart in electron

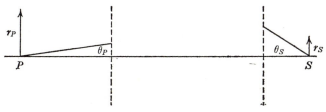

Fig. 3.4. Illustration of the Helmholtz–Lagrange theorem.

optics. In fig. 3.4 let there be any optical system in the space between the dotted lines and let this system be symmetrical about the axis PS. Rays from an object of height r_P at P pass through the system and are brought to a focus to form an image of height r_S at S. θ_P is the angle which a ray from P makes with the axis and θ_S is the angle which the same ray makes with the axis when it converges towards S. In the diagram relatively large angles are shown for the sake of clarity but we shall be concerned only with systems in which both angles are extremely small. The theorem can then be stated in the form

$$n_P r_P \theta_P = n_S r_S \theta_S, \tag{3.2}$$

where n_P and n_S are the refractive indices in the object and image spaces respectively. In electron optics it can be shown that the effective refractive index is proportional to the square root of the potential, measured from a point at which the electrons would be at rest, so (3.2) becomes

$$r_P \theta_P \sqrt{V_P} = r_S \theta_S \sqrt{V_S}. \tag{3.3}$$

We can now apply this result to the system of fig. 3.4, taking r_S to be the radius of a very small central area of the probe spot, over which the current density J_S can be considered uniform, and r_P the radius of the corresponding area of the crossover. θ_S is the semi-angle of the cone of rays which converge to form the probe spot and θ_P is the semi-angle of the cone containing these same rays when they leave the crossover. Let J_P and J_S be the current densities parallel to the axis, resulting from these rays, at the crossover and at the probe spot respectively. Since all rays are very nearly parallel to the axis, we may write from (3.3)

$$J_P/J_S = r_S^2/r_P^2 = V_P\theta_P^2/V_S\theta_S^2. \tag{3.4}$$

According to Langmuir's law the maximum possible current density J_S' in the probe spot is, from (2.2),

$$J_S' = J_0 e V_S \theta_S^2/kT \quad A/m^2, \tag{3.5}$$

and we now see from (3.4) that this requires a corresponding limiting current density J_P' in the crossover, where

$$J_P' = J_0 e V_P \theta_P^2/kT \quad A/m^2. \tag{3.6}$$

It is to be emphasized that J_P' is not the total current density in the crossover; it is the current density resulting from the cone of rays which can be accepted by the electron-optical system following the gun. This narrow cone includes a solid angle of $\pi\theta_P^2$, so (3.6) can be expressed in the form

$$\beta' = J_0 e V_P/\pi kT \quad A/m^2 \text{ steradian}. \tag{3.7}$$

In general, β is a measure of the useful current density in the crossover and is termed the *brightness*: it is expressed in amperes per square metre per steradian. β' is the maximum brightness permitted by Langmuir's law and the ratio of β to β', expressed as a percentage, is one measure of the efficiency of the gun. Another measure of efficiency, from a different point of view, is the ratio of the useful current projected into the dotted cone of figure 3.3 to the total current leaving the cathode. The design of the stabilized power unit which provides the accelerating voltage is greatly simplified if the total current is kept as small as possible.

Other characteristics of an electron gun which are important in scanning electron microscopy are the radius of the crossover, which determines the number of stages of demagnification that must follow the gun, and the life of the thermionic cathode.

3.1.3. Space charge

So far we have neglected any possible effects of space charge in the electron-optical system and these must now be considered. One obvious place where they might be important is near the surface of the cathode. Should space charge be significant in this region, a potential minimum will occur in front of the cathode and this will cause many of the emitted electrons to return to the cathode. The effective current density J_0 is then that which results from electrons emitted with sufficient energy to cross the potential barrier and this may be much smaller than the current density originally emitted from the cathode.

Since it is advantageous to make J_0 as large as possible, it is pertinent to enquire why space charge should ever be allowed to reduce its effective value. One reason is that certain types of thermionic emitter have short lives if operated under conditions of temperature limitation of emission; with these, the presence of space charge to reduce the electric field at the surface of the cathode is a practical necessity. When the current from the cathode is limited by space charge, the conditions depicted in fig. 3.1 are somewhat modified, since the lens action of the electrodes is perturbed by the field produced by space charge. As a result, no sharp image of the cathode surface is formed in the plane B, though gross features are usually visible (Moss, 1968, p. 37). However, an effective crossover still exists and the remainder of our discussion is still valid.

Even if a cathode can be used under conditions of temperature limitation of emission, there may be difficulty in producing at the cathode surface a sufficiently strong electric field to draw away the full saturated current. With a triode gun of the form shown in fig. 3.1, the minimum spacing between anode and grid is limited by voltage breakdown and further intensification of the field at the cathode surface can only be achieved by altering the form of the electrodes. In one successful structure, which we shall be considering later, a sharply pointed cathode is used and its tip is allowed to project through the hole in the grid (§ 3.1.8). Another attack on the problem has been made by Simpson and Kuyatt (1963), who were particularly concerned with guns in which the final energy of the electrons was quite low. They used more than three electrodes so that the electrons could be drawn from the cathode by a strong field and subsequently decelerated to the required energy. Guns of this type may find application in scanning microscopes if operation at voltages of the order of 1 kV is required.

So far, we have discussed the effects of space charge only in the region of the cathode. After leaving the crossover, a very narrow beam of electrons traverses a distance of the order of a metre and it might be thought that their mutual repulsion would be important: it is certainly a limiting factor in the design of cathode-ray tubes. This problem has been considered by a number of authors (Thompson and Headrick, 1940; Hollway, 1962; Schwartz, 1957) and it appears that, for the very small currents used in scanning microscopes, the effect of space charge in the beam is quite negligible.

3.1.4. Space-charge limited guns

Guns in which electron emission from the cathode is intentionally limited by space charge are not, at the present time, of much importance in scanning electron microscopy. We include them partly

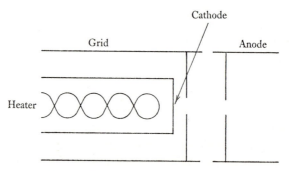

Fig. 3.5. The gun of a cathode-ray tube.

because they have been more intensively studied than any other type, but also because they may become important in the future. They are widely used in cathode-ray tubes, where they have the general form shown in fig. 3.5. The end of the cathode cylinder is coated with a mixture of alkaline–earth oxides and gives good emission at a temperature of about 1100 °K.

Such guns conform quite closely to the idealized structure visualized in (§3.1.1.), except for the limitation of emission by space charge, and a very complete study of their design and mode of operation has been made by Moss (1968). Moss states that guns of this type are normally operated at emission current densities in the range 1–10 A/m² and that any attempt to draw current densities in excess of 100 A/m² would be likely to lead to short life. In a properly designed cathode-ray tube the current density in the focused spot will probably be in the range from 0.4 to 0.7 of the maximum theoretical value predicted by Langmuir's equation. Failure to reach the full theoretical value may result from non-uniformity of emission, from aberrations in the triode lens or from electron interaction effects in the potential minimum in front of the cathode and/or in the crossover. This problem has been considered by Haine (1957).

That guns with cathodes of this type have not hitherto been used in electron microscopes can be ascribed to two facts: the emission current density is not as high as can be obtained with other cathodes and, in addition, oxide cathodes do not stand up well to the indifferent vacua commonly achieved in demountable apparatus and to frequent exposure to atmospheric pressure. These shortcomings are not necessarily insurmountable and guns of this type may, in the future, find application in scanning microscopy.

3.1.5. The tungsten-hairpin cathode

So far as electron microscopes are concerned, the most widely used form of electron gun is of the type shown in fig. 3.6. The cathode is a hairpin of tungsten wire heated by direct current. The diameter of the wire is usually about 0.125 mm, which gives a reasonable compromise between filament life, requiring thick wire, and minimum heating current, favouring thin wire. If the filament current is too great, dissipation of heat may present problems. The cathode is surrounded by a negatively-biased grid shield, of which two forms are shown in fig. 3.6 (a) and (b) respectively. The shape of the shield does not seem to be of great importance and we shall consider only the flat grid of fig. 3.6 (a). On the side of the grid remote from the filament is an anode to which the accelerating voltage is applied. Important parameters are the diameter d of the hole in the grid shield and the height h from the tip of the filament to the front surface of the flat grid. With grids of the form of fig. 3.6 (b), h may be negative.

Although guns of this type have been investigated by a number of research workers (Haine and Einstein, 1952; Dolder and Klemperer, 1957; Haine, Einstein and Borcherds, 1958; Boersch and Born, 1960) it cannot be said that there is complete agreement about their mode of operation. For example, there has been discussion as to whether a real crossover exists in such a gun or whether the electron-optical system of the microscope does not, in fact, image a portion of the tip of the tungsten filament. This is not really important;

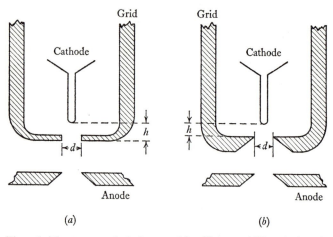

Fig. 3.6. The tungsten-hairpin gun. After Haine and Einstein (1952).

when electrons pass from the gun into field-free space they will move in straight lines and, if these lines are produced backwards, they will appear to originate in an area which we may term the virtual object. Whether this virtual object is a true crossover or a portion of the cathode surface is immaterial and we shall continue to refer to it as the crossover. Again, opinions differ about the effects of space charge and of aberrations in the lens formed by the triode electrodes, but these are matters which affect the internal operation of the gun rather than its external characteristics, with which we shall be chiefly concerned. About these there is no serious disagreement and we shall summarize the results that have been obtained, relying largely on data taken from the paper by Haine and Einstein (1952).

The way in which four important quantities vary with grid bias is shown in fig. 3.7 for a representative gun, with plane grid,

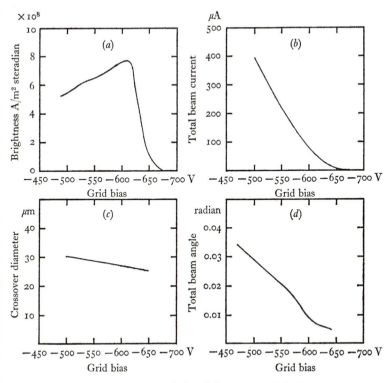

Fig. 3.7. Characteristics of the tungsten-hairpin gun.
After Haine and Einstein (1952).

fig. 3.6 (*a*), in which $d = 1.75$ mm, the accelerating voltage was 50 kV and the tungsten filament was run at a temperature of 2650 °K. The life of a tungsten filament is not appreciably reduced if full saturation emission is drawn from it; hence it is advantageous to operate the gun so that this condition is fulfilled and the maximum brightness thus obtained. We might then hope that this maximum would approach the limit set by Langmuir's law. From fig. 3.7 we see that the brightness, for a narrow cone of electrons moving nearly parallel to the axis, varies markedly with grid bias and attains a maximum at a bias rather less than the cut-off value. Within the limits of experimental error, this maximum was found to be equal to the theoretical Langmuir limit. In general it was found that, over a range of values of d and h and with some variation of the shape of the grid electrode, the theoretical brightness could be attained if the cathode temperature did not exceed 2700 °K and if the gun was operated at the correct bias. For temperatures exceeding 2700 °K, the increased emission caused space-charge effects to become important and the maximum brightness fell below the Langmuir limit, reaching about forty per cent of the limit at 2900 °K.

The curves of fig. 3.7 (*b*) and (*d*) show what happens when the negative grid bias is reduced below its optimum value. More current is drawn from the cathode, but it is distributed over a beam of wider angle. There is a further effect, which the curves do not show but which becomes apparent if a magnified image of the crossover is obtained on a fluorescent screen. Whereas the image is a roughly circular disc when the gun is adjusted for optimum brightness, at smaller negative values of bias the shape of the image becomes more irregular and may break up into a central patch surrounded by a bright ring. Effects of this kind, which are quite sensitive to changes in the geometrical structure of the gun, arise because the electron emission comes from a tungsten hairpin. For bias near the cut-off value, only the tip of the hairpin experiences a field drawing electrons away from its surface and the effective cathode then approximates to a portion of a sphere. Under this condition the crossover is circular. As the negative bias is reduced, current is drawn from an ever increasing portion of the hairpin and, not surprisingly, the crossover assumes various irregular shapes. The above discussion shows that it is of the first importance that the gun should be operated with the correct bias.

The curve of fig. 3.7 (*c*) shows that the diameter of the crossover is not greatly affected by moderate variations of grid bias in the region of the optimum bias. It is also found not to be very sensitive to changes in gun geometry, so long as the cathode shape and wire diameter remain constant. Turning now to variations in d and h, an increase in d allows the field produced by the anode to penetrate more readily into the space within the grid so that, for a given value of h, the negative cut-off voltage is increased. Similarly, for a given value of d, an increase in h reduces the negative cut-off voltage. With an accelerating voltage of 50 kV, satisfactory values of d might lie in the range from 0.75 to 2.25 mm, with corresponding values of h in the range from 0.25 to 1.25 mm.

To extend these results to guns operated at lower voltages we might seek to use scaling theory, which is discussed in some detail by Moss (1968, p. 191). It is a consequence of the basic equations of motion of electrons in electrostatic fields that, if all rectangular components of the field are scaled in the same ratio and if both space charge and initial velocities are negligible, the shapes of the electron trajectories are unchanged. The scaling can be brought about either by altering all linear dimensions of electrodes in the same ratio, or by changing all voltages in the same ratio, or by altering both simultaneously. Since Haine and Einstein have shown that performance is beginning to be limited by space charge even in guns working at 50 kV, it is obviously desirable to avoid any reduction in electric field strength when operation at lower voltages is contemplated. Thus linear dimensions of electrodes ought to be scaled down in a ratio not less than that of the voltages, but this would involve a decrease in filament diameter leading to an unacceptable reduction of life. We should expect the ratios of filament diameter to the quantities h and d to be of importance in determining gun efficiency and, as is shown by the figures given above for a 50 kV gun, there is considerable latitude in the choice of these ratios. It would therefore seem reasonable, when designing a gun for operation at lower voltage, to take the maximum values of h and d recommended for 50 kV (say $h = 1.25$ mm, $d = 2.25$ mm) and to scale these down in proportion to the voltage, while leaving the filament diameter unaltered. For operating voltages down to about 15 kV this would still leave the ratios of filament diameter to h and d within the limits found satisfactory at 50 kV by Haine and Einstein. In the absence of published experimental data on operation at the lower voltages, we cannot be sure that this design procedure will produce the best gun.

At still lower voltages, space charge is likely to become more important because it will be mechanically impracticable to scale down electrode dimensions and spacings to compensate for the reduction in voltage. Dolder and Klemperer (1957) have published some experimental results on guns operating at voltages as low as 4 kV. They conclude that the distance from cathode to grid should be about three-quarters of the radius of the grid aperture and that, under these conditions, the brightness will be of the order of fifty per cent of the Langmuir theoretical value. These authors were concerned with guns giving electron beams of considerably larger angle than those studied by Haine and Einstein, so the two investigations are not strictly comparable. This is illustrated by the fact that Dolder and Klemperer, working with an accelerating voltage of 4 kV, used a grid aperture of diameter 2 mm; this is roughly equal to the maximum diameter used by Haine and Einstein at 50 kV. Further experimental work on low-voltage guns would be welcome, but perhaps we may draw the general conclusion that, for voltages in excess of 4 kV, it is possible to design guns whose performance approaches the theoretical limit.

The importance of operating the gun with correct grid bias has already been emphasized and, when the bias voltage is supplied from a stabilized power unit, this presents no difficulty. However, the anode is normally kept at earth potential while the cathode is at a high negative potential, so it is often convenient to obtain the bias from the voltage drop across a resistor in series with the cathode, fig. 3.8, and thus to avoid the provision of a separate supply at a high potential with respect to earth. When this is done, the bias voltage V_b is a function not only of the resistance R, but also of the accelerating voltage, the geometry of the gun and the cathode temperature. To obtain proper operating conditions, R should therefore be variable. This problem has been considered in greater detail by Haine, Einstein and Borcherds (1958) but, since they were concerned with accelerating voltages of 50 kV, their results are not directly applicable to the lower-voltage guns used in scanning microscopy.

To obtain the maximum brightness from a gun with a tungsten-hairpin cathode the filament must be run at as high a temperature as possible but, beyond a certain point, evaporation of tungsten reduces the life of the filament to an unacceptable value. The life may also be reduced by chemical attack by residual gas in the vacuum system; chiefly oxygen and water vapour. These matters have been investigated by Bloomer (1957a and 1957b), whose results are quoted below.

The cause of burn-out of a filament can readily be identified by inspection, since attack by residual gas causes a fairly uniform reduction of cross-section over the whole length of the hot part of the filament; it is not greatly affected by filament temperature. Evaporation, on the other hand, increases rapidly with temperature and has its greatest effect at the hottest point of the filament, which usually lies just to one side of the tip of the hairpin. The reduction in cross-section in the vicinity of this point causes a local rise in temperature, which increases evaporation and results in a local burn-out. Because the two causes of burn-out are affected differently by temperature, chemical attack will be the more important at

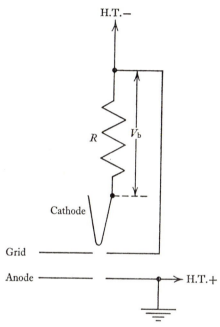

Fig. 3.8. The provision of bias for the tungsten-hairpin gun.

low temperatures, while evaporation will predominate at high temperatures.

Bloomer found that the life of a filament ended when the diameter had been reduced by about six per cent. His results on the relation between life and temperature for a filament of 0.125 mm diameter are summarized in the curves of fig. 3.9 where the full-line curve refers to life terminated by evaporation alone, while the dotted curves indicate the effects of various pressures of residual gas in the absence of evaporation. In practice, the gas pressure will usually be less than 10^{-4} Torr, so evaporation is likely to be the controlling factor. When considering the best practical operating temperature

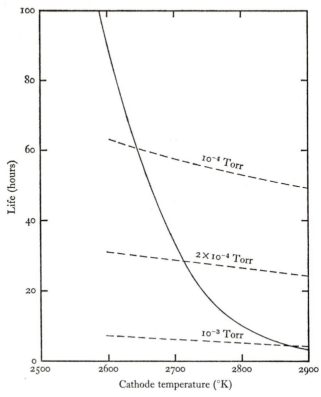

Fig. 3.9. Variation of the life of tungsten filament with temperature and ambient gas pressure. After Bloomer (1957b).

we recall that, at temperatures above 2700 °K, Haine and Einstein found the effects of space charge to be appreciable, so that brightness did not increase in proportion to the density of electron emission. At the higher temperatures, also, life rapidly becomes unacceptably short, so operation in the region of 2700 °K, with a life of about thirty hours, would seem to be a reasonable compromise. For special purposes, a somewhat higher temperature might be worth while.

One further complication should perhaps be mentioned. Heating power for the filament usually comes from either a constant-current or a constant-voltage supply. If the former, the filament temperature will tend to rise during life, as its diameter is reduced by evaporation. With a constant-voltage supply, the temperature will tend to fall during life. The magnitude of the latter effect will be modified by any added resistance that may be present in the filament circuit.

3.1.6. Field emission and T-F emission

It will be clear from the previous section that the current density obtainable by thermionic emission from a tungsten filament is a good deal lower than one would like and that an acceptable value is achieved only at the expense of filament life. It has been known for many years that much larger current densities can be drawn from a metal when the surface of the latter is subjected to an intense electric field directed so as to pull the electrons out of the metal. This overall result arises from three separate effects, as follows.

In general, the electrons in a metal are prevented from escaping by the existence of a potential barrier at the surface but, at any temperature above the absolute zero, some electrons will have sufficient energy to surmount this barrier and so to escape. This is the phenomenon of pure thermionic emission. For tungstem, which is the material with which we shall be principally concerned, the height of the potential barrier is about 4.5 eV and thermionic emission is negligible below about 1500 °K: as we have already seen, temperatures in the region of 2700 °K are needed to secure sufficient emission for electron-microscopical purposes. If the surface of the emitter is subjected to an accelerating field F V/m, the effective potential barrier is lowered and the emission increases. This is the effect first explained by Schottky; it results in an increase of emission by a factor of $\exp[0.44(\sqrt{F})/T]$. For a field of 10^8 V/m, the factor would be about five. At still higher fields, of the order of 10^9 V/m, another effect comes into play because the field not only reduces the height but also the width of the surface potential barrier. It is then possible for electrons to pass through the barrier by quantum-mechanical tunnelling, instead of having to acquire sufficient energy to climb over the barrier. This effect was first explained by Fowler and Nordhein, whose work leads to the following relation between current density and field strength when the work function of the metal is ϕ,

$$J = AF^2 \exp(-B\phi^{\frac{3}{2}}/F), \tag{3.8}$$

where A and B are approximately constant though each depends to some extent on the work function and B also on the field strength.

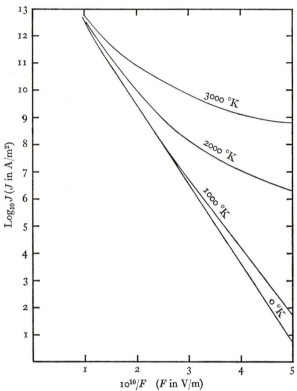

Fig. 3.10. Variation of the current density of emission with temperature and field strength. After Dyke and Dolan. In *Adv. inelec. and electron phys.* ed. Marton **8**, 89. © Academic Press, 1956.

It is to be noted that this effect is independent of temperature, so long as ϕ does not vary.

When the cathode is at a high temperature and is subject to a strong accelerating field, all three of the above effects may contribute to the total emission and the overall result for a tungsten cathode is shown in fig. 3.10. The curve for $0\,°K$ represents the limiting case where the tunnelling mechanism accounts for the whole of the observed emission; the curve for room temperature would not be sensibly different. When this condition obtains we shall speak of *field emission*. At higher temperatures, where thermionic emission is appreciable but where the majority of the electrons nevertheless escape by the tunnelling process, it is customary to refer to *T-F emission* (i.e. thermionic-field emission). Both types of emission have been extensively studied and there is good agreement between theory and experiment up to current densities of 2×10^{12} A/m². Excellent reviews of the subject have been given by Dyke and Dolan (1956) and by Gomer (1961).

In view of the very high current densities that can be obtained in the presence of strong fields, it is natural to enquire whether the phenomena that we have been discussing can be used in the design of electron guns. In this section we shall consider certain general matters that are relevant to this question; the actual design of guns will be dealt with later.

To obtain reasonable enhancement of emission from the application of a field, field strengths in excess of 10^9 V/m are needed. To

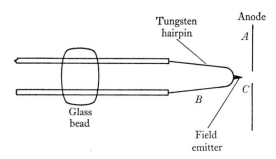

Fig. 3.11. Construction of a field-emission cathode.

produce such fields with reasonable voltages, a very sharp needle-shaped cathode is used, the radius of curvature of the tip being normally less than $1.0 \, \mu$m. Although a great many possible cathode materials have been and are being investigated, tungsten has generally been found most suitable on account of its mechanical strength and high melting point. A common method of construction is to weld about 1 mm of tungsten wire C, in fig. 3.11, to a loop of tungsten wire B. The end of C remote from B is reduced to a fine point by electrolytic etching and the system is mounted in front of an electrode A which can be maintained at an appropriate positive potential with respect to C. The temperature of C can be raised by passing current through the tungsten loop; this permits the emitter to be outgassed even if subsequent measurements are to be made, with C at room temperature.

Field emission is sensitive to crystal structure and some crystal planes in tungsten give much more copious emission than others under similar conditions. Efforts are usually made to secure favourable orientation of the planes in the emitter either by using single-crystal wire or by subjecting the emitter to prolonged pre-heating to encourage the growth of large crystallites. The emission is also greatly affected by residual gas unless an extremely high vacuum is maintained. Gas molecules become adsorbed on the surface of the emitter and, by altering its work function, change the emission. The molecules also become ionized and the positive ions so formed bombard the cathode. The sputtering which results can cause large changes in the local radius of curvature of the emitter and hence in the field to which it is subjected.

The conditions under which stable cold field emission can be obtained have been studied by Martin, Trolan and Dyke (1960). In effect they find that the ambient pressure must at all times be of the order of 10^{-12} Torr and that, if this condition is met, stable continuous current densities up to 10^{11} A/m^2 can be drawn from the cathode.

When the emitter is raised to a temperature of the order of 1000 °K, a number of interesting effects result. These have been studied by Dyke, Charbonnier, Strayer, Floyd, Barbour and Trolan (1960). In the first place, adsorbed gas molecules do not remain on the emitter for any length of time, so a relatively clean surface can be maintained. From this point of view, the hotter the emitter, the cleaner its surface will be. It is also found that, at temperatures well below that at which appreciable vaporization takes place, surface atoms of tungsten become mobile and, under the influence of surface tension, tend to move away from the regions of smallest radius of curvature. In this way, irregularities resulting from positive-ion bombardment are repaired and the surface of the emitter remains smooth. Thus the raising of the emitter to a high temperature largely overcomes two of the major difficulties encountered with cold field emission. There are, however, further effects.

The motion of tungsten atoms under the influence of surface tension not only smoothes out the irregularities caused by positive-ion bombardment, it also causes the normal radius of curvature of the tip of the emitter to get progressively larger. This can be a useful effect when the emitter is first made, since it enables one to start with a point which is sharper than required and then to bring the radius of the tip to the appropriate value by heating. However, the change of radius does not stop at the required value though, as one might expect, the rate of change gets slower as the radius increases. Eventually a radius is reached at which the shape of the tip is relatively stable, but this radius does not necessarily correspond to the field that one wishes to apply. Fortunately, there is an opposing effect in that the electric field itself opposes the surface tension and tends to cause the migrating tungsten atoms to move towards the tip. Motion of the atoms away from the tip, under the action of surface tension alone, is often known as 'dulling', while migration in the opposite direction caused by the field is referred to as 'build-up'. By suitable choice of the initial tip radius, the field and the temperature, it is possible to construct a system in which the tip radius does not vary appreciably with time. Alternatively, if the emitter is used to give pulsed emission, the duty-cycle and the temperature can be chosen in such a way that build-up while the field is applied is just balanced by dulling while the field is zero. In either case, stable emission can be obtained with a residual gas pressure of the order of 10^{-6} Torr.

We might hope that, if the conditions discussed above were fulfilled, the emitters would have very long lives. With a cold tungsten field emitter operated with a residual pressure of 10^{-12} Torr, this hope has been realized and a life of several thousand hours can be expected. At somewhat higher pressures, ion bombardment and surface adsorption seriously reduce the stable life, but the emitter can be periodically re-conditioned by a brief flash heating which cleans the surface and repairs the bombardment damage. Using this procedure, useful operation can be obtained at residual pressures as high as 10^{-9} Torr. When T-F emission is used, the life of the cathode is usually brought to an end as the result of a vacuum arc and this, in turn, appears to be caused by the diffusion of impurities to the surface of the tungsten tip. For long life, therefore, the purest possible wire should be used and, since the rate of diffusion increases rapidly with temperature, this should not be too high. On the other hand we have seen that a high temperature is desirable to remove adsorbed gas from the surface, so a compromise must be reached. With the purest tungsten wire at present obtainable, the optimum temperature is slightly above 1000 °K giving lives of a few hundred hours with a residual gas pressure of 10^{-6} Torr, when emission current densities up to 10^{11} A/m² are drawn. Looking to the future, wire of still higher purity, operated at a somewhat higher temperature, would be expected to give longer lives.

It is clear that, under suitable circumstances, either cold field emission or T-F emission can provide current densities about a million times as great as those obtainable thermionically from a tungsten hairpin. It might therefore be thought that the use of a field emitter in an electron gun would lead to an increase of brightness by the same factor. This, however, is not the case, as was shown by Drechsler, Cosslett and Nixon (1958). We give below a simplified treatment of the problem which leads to the same conclusions as those reached by these authors.

In a gun using a thermionic cathode such as a tungsten hairpin, the source of electrons is of relatively large size and the final probe is obtained by successive demagnification with two or three lenses. On the other hand, when field emission is employed, the emitter must necessarily be very sharply pointed to give the requisite electrostatic field. Moreover, because this field is so strong, electrons leave the tip of the emitter in directions which are very nearly normal to the surface and the effective size of the source is almost negligibly small. Further demagnification is unnecessary and the diameter of the final probe is determined almost entirely by spherical aberration in the focusing system. This situation is represented in fig. 3.12 where r is the radius of curvature of the tip of the emitter and, to avoid complication, the focusing system is represented as a single thin lens L. S_1 and S_2 are the distances from L of the effective source and the probe respectively.

If the diameter d of the probe spot is determined entirely by spherical aberration, we shall later show (§ 3.2.3) that

$$d = \tfrac{1}{2}C_s\alpha_2^3, \tag{3.9}$$

where C_s is the spherical aberration coefficient.

Also
$$S_1\alpha_1 = S_2\alpha_2. \tag{3.10}$$

In practice both α_1 and α_2 are very small so that, if J_F is the current density of emission from the tip, the current I_F in the final probe is given by

$$I_F = \pi r^2 \alpha_1^2 J_F = \pi r^2 J_F S_2^2 \alpha_2^2 / S_1^2 \qquad (3.11)$$

or substituting for α_2 from (3.9),

$$I_F = \pi r^2 J_F S_2^2 (2d)^{\frac{2}{3}} / S_1^2 C_s^{\frac{2}{3}}. \qquad (3.12)$$

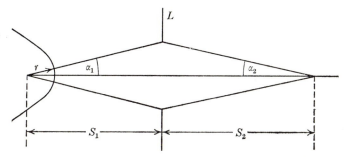

Fig. 3.12. The electron-optics of a field-emission cathode.

It has already been shown (2.36) that, when a thermionic cathode is used, the maximum current I_T in the probe is given approximately by

$$I_T = 7 \times 10^3 d^{\frac{8}{3}} J_0 V / C_s^{\frac{2}{3}} T, \qquad (3.13)$$

where J_T is the current density of emission from the cathode, V the accelerating voltage and T the temperature of the cathode.

When comparing (3.12) and (3.13) it must be remembered that (3.12) is no more than a rough approximation. The actual focusing system will consist of a combination of the gun proper and any lens which follows the gun, and will certainly not be equivalent to a single thin lens. Furthermore we have neglected chromatic aberration which may well play a part in determining spot size, though spherical aberration is likely to be dominant. Finally, the focusing system of a field-emission gun is likely to be quite different from that of a thermionic gun, so C_s will be different in the two cases, though it is unlikely to be vastly different. For our present purpose we are interested only in an order-of-magnitude calculation, so we ignore these uncertainties and assume C_s to be the same in the two cases. Dividing (3.12) by (3.13), we then have

$$\frac{I_F}{I_T} = 7.15 \times 10^{-4} \left(\frac{J_F}{J_T}\right) \left(\frac{S_2}{S_1}\right)^2 \left(\frac{T}{V}\right) \left(\frac{r}{d}\right)^2. \qquad 3.14$$

For the scanning microscope we may take as typical values

$$J_{\mathrm{F}}/J_{\mathrm{T}} = 10^6, \quad S_2^2/S_1^2 = 0.1, \quad V = 3 \times 10^4 \text{ volts},$$

$$T = 2700 \,^{\circ}\text{K}.$$

We then have $\qquad\qquad I_{\mathrm{F}}/I_{\mathrm{T}} = 13r^2/d^2.$ $\qquad\qquad$ (3.15)

Despite the uncertainties of our calculation, the following points emerge quite clearly. The field-emission gun will be superior to the thermionic gun only if d/r is sufficiently small. In one field-emission gun to be described later (§ 3.1.7) r was made equal to about 0.06 μm and (3.15) then indicates that this gun would give greater probe current than a thermionic gun for spot diameters less than about 0.2 μm; for larger diameters the thermionic gun would have the advantage. When small probe diameters are in question, the superiority of the field-emission gun is very marked. For example, if $d = 0.01 \,\mu$m, (3.15) gives $I_{\mathrm{F}}/I_{\mathrm{T}} = 467$ for the gun mentioned above.

It should be re-emphasized that relatively little reliance can be placed in the numerical factors obtained in the above equations. Apart from the uncertainties already mentioned, further operational experience is needed before one can state with confidence the value of current density to be assumed for a field-emission cathode. Although values as high as 10^{11} A/m have been reported when the best possible vacuum conditions are provided, it does not follow that equally good results will be obtained consistently under the operating conditions of a scanning microscope, where facilities for outgassing may be restricted.

Further work is also needed to determine the best design of field-emission gun. The above discussion has shown that conditions in such guns are quite different from those in thermionic guns, so the optimum disposition of electrodes is likely to be different also. Theoretical analyses of certain aspects of this problem have been given by Everhart (1967) and by Worster (1969).

So far, nothing has been said about the spread of energy in the electrons emitted from a field-emission cathode. This spread will depend on the field at the cathode surface and on the temperature of the cathode, but will normally be about one electronvolt. It is therefore not very different from that obtained with a thermionic cathode.

3.1.7. Field-emission and T-F emission guns

A detailed account of work on the design and operation of a cold
field-emission gun has been given by Crewe, Eggenberger, Wall and
Welter (1968) and a schematic diagram of their gun is shown in
fig. 3.13. There are two anodes and the positive voltage V_1, applied
to the first, controls the emission current, while V_0 determines the
final energy of the electron stream. The electrode shapes result from
a theoretical study by Butler (1966) and are arranged to give
negligible field in the vicinity of the apertures. This, together with
the correct form of field between the two anodes, reduces spherical
aberration to a minimum.

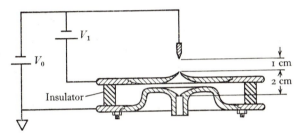

Fig. 3.13. The field-emission gun. Redrawn by permission from
Crewe, Eggenberger, Wall and Welter (1968).

The emitter tip was made from 0.125 mm tungsten wire welded
to a tungsten hairpin so that it could be heated when necessary.
Special wire was used for the tip in which the (310) planes were
oriented at right angles to the axis; this orientation gives intense
emission along the axis. The initial shaping of the tip was carried
out by electrolytic etching in sodium hydroxide solution and it was
then formed by heat treatment in a vacuum. In this way a tip radius
of 0.1–0.2 μm was obtained but, for a reason which will shortly be
explained, this was considered to be too large. The tip was therefore
subjected to a remodelling process described by Sokolovskaia
(1956), in which pulses of heating current were passed through the
hairpin while a positive voltage of 3–8 kV with respect to the anode
was applied to the tip. Under this treatment with reverse bias the
tip radius could be reduced to about 0.06 μm and it was then possible to obtain the requisite emission using a voltage V_1 of 1–2 kV

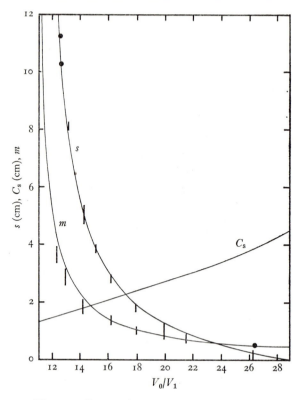

Fig. 3.14. Characteristics of the field-emission gun. After
Crewe, Eggenberger, Wall and Welter (1968).

between the tip and the first anode of the gun. Other workers have
not found this re-modelling process to be essential.

The performance of the gun is shown in fig. 3.14, where the curves
represent Butler's theoretical predictions and the points indicate the
experimental results so far obtained. The curves show that both the
magnification m and the image distance s increase very rapidly as
the ratio V_0/V_1 is reduced. Thus it will often be desirable to keep V_1
below 2 kV and this explains why remodelling of the tip to reduce its
radius to about 0.06 μm may be necessary.

The gun described above has been used in a scanning microscope
where the pressure was maintained at about 10^{-9} Torr and under
this condition, appears to be giving fairly satisfactory performance.
When the emission becomes erratic, the emitter tip can be re-formed

by the treatments previously discussed and the time between these treatments depends principally on the pressure. At pressures below 10^{-9} Torr it may be several tens of hours. Theoretically the gun should be able to produce a focused spot with a radius of $2.5 \times 10^{-3} \mu$m or less, the value depending on the ratio of V_0 to V_1 so that no additional lens is needed to provide further demagnification. With V_0/V_1 being set equal to twelve, the expected spot radius was $10^{-2} \mu$m and an experimental value of $2.5 \times 10^{-2} \mu$m was recorded. Again, theory indicates that, at radii of this order, the brightness should be between one hundred and one thousand times as great as that which can be obtained with tungsten-hairpin thermionic guns. Experimental confirmation of this prediction is at an early stage, but there seems little doubt that the field-emission gun is very much brighter than a tungsten thermionic gun.

When we turn from cold field emission to T-F emission, the experimental evidence is far from clear. Guns operating with hot tungsten-point cathodes have been reported by several investigators, e.g. Hibi (1956), Sakaki and Möllenstedt (1956), Maruse and Sakaki (1958), Swift and Nixon (1962), Speidel (1965), and Hanszen and Lauer (1967a). In some of these investigations the point cathode was substituted for a tungsten hairpin in the standard gun structure of a transmission electron microscope, and the purpose was to obtain a very small effective source rather than improved life or greater brightness. In such cases the residual pressure might have been as high as 10^{-4} Torr and the field at the cathode surface was usually too small to produce appreciable field emission. In no instance was any attempt made to satisfy the conditions stipulated by Dyke, Charbonnier, Strayer, Floyd, Barbour and Trolan (1960) for a balance between the surface-tension forces and the field forces acting on the mobile tungsten atoms at the emitter surface. In one case, an early claim of very high brightness was subsequently withdrawn. At the time of writing, it must be concluded that there is no reliable record of a T-F gun having given a performance in respect of life or brightness which is appreciably better than that to be expected from an ordinary thermionic gun with a tungsten hairpin cathode. Further investigation is clearly needed.

3.1.8. Other guns with thermionic cathodes

We have seen that guns using a tungsten-hairpin cathode, although widely used, suffer from a number of disadvantages, as follows:

(*a*) The maximum current density of emission that can be obtained is a great deal lower than one would like and, as a result, scanning times in the microscope are undesirably long.

(*b*) Even if the emission current density is limited to the lowest tolerable value, the filament life is unlikely to exceed a few tens of hours.

(*c*) The temperature of the emitter is very high and this reduces the effective brightness of the gun which depends on the ratio of emission density to temperature.

(*d*) If the temperature of the filament exceeds about 2700 °K, space charge begins to limit the current that can be drawn from the gun.

In the search for more satisfactory emitting materials, a good deal of attention has been paid during the past decade to the borides of the alkaline-earth metals and of lanthanum, cerium and thorium, all of which form compounds of the type MB_6. The thermionic properties of these compounds have been investigated by Lafferty (1951), who found lanthanum hexaboride to be the most useful member of the group. He concluded that the active emitting material was a layer of lanthanum, maintained on the surface of a highly refractory cage of boron atoms and continually replenished by diffusion from the interior of the cathode. However, more recent work by Ahmed and Broers (1972) does not substantiate this view. Whatever may be the true facts, there is no doubt that lanthanum hexaboride can give emission current densities of the order of 5×10^5 A/m² at an operating temperature of about 2100 °K, with a life of some 100 hours; a very much better performance than one can obtain with tungsten. No forming process is needed to obtain emission, other than heating the cathode to its operating temperature for a few minutes. The current density usually increases during the first few hours of life. Operation is not impaired by the residual gas in a demountable system at pressures of the order of 10^{-6} Torr and, after exposure to the atmosphere, the cathode can be reactivated by heating to its normal running temperature for a few minutes.

A major difficulty in the use of lanthanum hexaboride as a cathode material arises from its very high reactivity, at the operating temperature, with almost all materials on which it might be mounted. Moreover, at the present time, cathodes cannot be fabricated in a form that would permit direct heating by the passage of current. Details of a gun that successfully overcomes this difficulty have been published by Broers (1969 a); it takes the form shown in fig. 3.15. The cathode is a rod of lanthanum hexaboride, of 1 mm square

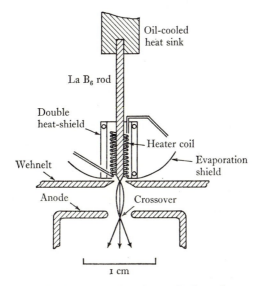

Fig. 3.15. The lanthanum hexaboride gun. Redrawn by permission from Broers (1969 a).

cross-section and 1.6 cm length. The rod is supported from an oil-cooled heat sink, into which one end is brazed. The other end is heated, partly by radiation from a surrounding coil of tungsten wire and partly by bombardment with electrons emitted by this coil. To make use of the very high current density emitted by the rod, a strong electric field must be created to draw the electrons away and thus prevent limitation by space charge. This is achieved by grinding the end of the rod to a fine point, with a radius of about 10 μm, and by allowing this point to project into the grid aperture. The original paper gives full details of the construction of the gun and of its measured performance. Guns of this type have now been in operation for a sufficiently long time to confirm their reliability and the excellence of their characteristics.

Another line of investigation in the search for improved cathodes has been pursued by Beck, Maloney and Mead (1970). These authors have been working with impregnated cathodes of the type developed for use in microwave valves. Such emitters consist of an indirectly-heated porous plug of tungsten or nickel, that has been impregnated with one or more alkaline-earth salts, which are subsequently activated by heat treatment. Sometimes a small percentage of a reducing agent, such as zirconium hydride is added. The methods of fabrication and properties of these cathodes have been reviewed by Beck (1959). They are very much more rugged than the earlier oxide-coated cathodes, can be operated at higher current densities and can be reactivated a number of times after exposure to the atmosphere. Recent improvements have led to emission densities of the order of $5 \times 10^5 \, A/m^2$. In this respect, therefore, they are comparable with lanthanum hexaboride, but with the impregnated cathodes the emission is obtained at the lower temperature of about $1500 \, ^\circ K$. Problems of overcoming space charge will be similar with the two types of cathode and it has been shown that impregnated cathodes can be fabricated with very small emitting areas, if this should prove necessary. At the present time there are no operational data on guns with impregnated cathodes, under the practical conditions obtaining in electron microscopes. Until information is available on the lives of these emitters and on their susceptibility to poisoning, one cannot predict how useful they will be. If they prove to be equally satisfactory in other respects, the fact that they can be indirectly heated without electron bombardment, will give them a great advantage over lanthanum hexaboride.

To conclude this section, mention must be made of work described by Albert, Atta and Gabor (1967). These authors have investigated emitters consisting of tungsten wire coated with a mixture of tungsten powder, metallic thorium and a small addition of zirconium. The thorium may be replaced by lanthanum and there is advantage in using tungsten carbide powder instead of tungsten. Emission current densities of about $10^5 \, A/m^2$ have been reported and still higher values can be obtained if the zirconium in the mixture is replaced by hafnium. The cathodes are not harmed by repeated exposure to the atmosphere or by the residual gas in a demountable system at a pressure of the order of 10^{-6} Torr. They would seem to be very suitable for use in guns for scanning microscopes but, at the time of writing, they do not appear to have been used for this purpose and further experimental results must be awaited.

3.1.9. General survey of electron guns

The electron gun has been dealt with at considerable length because, in some respects, it is the least satisfactory component of present-day scanning microscopes. The thermionic gun with tungsten-hairpin cathode is commonly used and, because of its great simplicity, there is no indication that it will be displaced, in the near future, on any large scale; for many purposes it is reasonably satisfactory. However, a cathode with longer life would be very acceptable and, for work at high resolution, the tungsten hairpin gives inadequate brightness. Higher brightness is also needed in applications where, for any reason, the total time of scan must be kept as short as possible.

The lanthanum hexaboride gun described by Broers offers a ten-fold improvement in brightness with a longer life than the tungsten hairpin, but at the cost of considerable additional complication and expense. Other thermionic emitters show promise, but have not yet been proved.

Cathodes employing field emission or T-F emission are potentially able to give very high current densities but, as we have shown, they offer little advantage over the tungsten hairpin unless a spot diameter less than about 0.2 μm is required. If field emission is compared with the lanthanum hexaboride gun, the change-over diameter is correspondingly smaller. Moreover, the emission that can be reliably obtained from a field-emission cathode, under the practical conditions obtaining in a scanning microscope, may well fall short of what is theoretically possible. Taking everything into account, if seems rather unlikely that field-emission cathodes will be used very extensively, though they will probably find application for special purposes.

3.2. Electron lenses

3.2.1. Introduction

Much of the early work on scanning electron microscopes was carried out with instruments employing electrostatic lenses. At that time it was not expected that resolution better than about 0.03 μm would be achieved and the electrostatic type of lens seemed adequate for the purpose in hand. Moreover, such lenses are somewhat cheaper to construct than magnetic lenses and they make less demand on the stability of the high-voltage power supply. Since the early days, resolution has steadily improved and experience has shown that magnetic lenses have marked advantages over the electrostatic variety. The aberration coefficients of the former are several times smaller than those of the latter and magnetic lenses are more easily dismantled for cleaning when this becomes necessary. Moreover, the strong electric field in an electrostatic lens tends to attract contaminating particles to the electrodes so that cleaning must be undertaken more frequently. At the present time electrostatic lenses are very rarely employed in scanning microscopes and we shall not give further consideration to them. Their use should not, perhaps, be completely ruled out for special-purpose instruments and, when necessary, suitable designs can be selected from data given in papers by Archard (1956) and by Hanszen and Lauer (1967*b*).

We shall be concerned with magnetic lenses which produce a field having symmetry about an axis and, in principle, any such field will have the properties of an electron lens. Axially-symmetric fields can be produced by air-cored coils without the use of any ferromagnetic material and we shall later (§3.2.8) consider a special case where a lens of this type may find application in a scanning electron microscope. For most purposes, however, lenses with iron cores and shields are preferred and we shall deal with these first. The essential features of a lens of this type are shown in fig. 3.16(*a*), where the whole structure has axial symmetry about the line *OZ*. A winding *B*, consisting of *N* turns carrying current *I* amperes causes magnetic flux to flow round the magnetic circuit formed by the polar cores *MM*, the end plates *EE* and the outer envelope *G*, all of which are made of a high grade of soft iron. The faces *FF* of the polar cores are separated by a distance S and a strong magnetic field is produced in the air gap so formed. It is here that the lens action occurs. The cores are drilled with axial holes of diameter D and the electron beam passes through these holes. The variation of flux density along the axis, in the vicinity of the air gap, is shown in fig. 3.16(c).

The simple arrangement described above may be varied in a number of ways to achieve particular ends and some of the possible modifications are shown in fig. 3.16(b). Here, the air gap has been

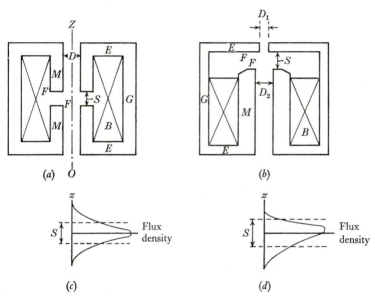

Fig. 3.16. The essential features of a magnetic lens.

moved from the centre to one end of the lens. This makes no difference to the action of the lens and may be convenient for other reasons. The edges of one pole piece have been bevelled and we shall later see that this may be desirable to avoid saturation of the iron. Finally, the bores in the two pole pieces are of different diameters, a variation which may be dictated by mechanical or electrical considerations. In this case the lens is said to be asymmetrical since, as shown in fig. 3.16(d), the variation of axial flux density in the vicinity of the air gap is now asymmetrical.

If the magnetic circuit of the lens is to operate efficiently, the relative permeability of the iron must be high and saturation is to be avoided. In practice, the relative permeability is likely to exceed 1000 if the flux density is not greater than about 1.25 Wb/m² and the cross-sectional areas at various parts of the circuit must be chosen to secure this end. This is indicated in fig. 3.16(*b*), where the thickness of material in the end plates and in the outer casing is less than the radial thickness of the polar core. However, a further point arises in the design of the core itself. There is a large magnetic potential difference between the faces *FF* of the two poles and thus an appreciable quantity of leakage flux enters the cylindrical surface of the core *M*. The total flux at the base of the core is therefore greater than that near the air gap and, if the latter is pushed to the limit, saturation may occur at the base. A simple increase in the diameter of a cylindrical core will not cure the trouble, since this causes a corresponding increase in the total flux without reducing the flux density. One solution of this problem is shown in fig. 3.16(*b*) and has been considered by Mulvey (1952). The radius of the lower pole piece is increased, but the air gap is made progressively larger by bevelling the pole face, so that the added iron does not produce a proportional increase in total flux. By proper choice of dimensions, saturation can be avoided. Care must also be taken that saturation does not occur at the pole face in the vicinity of the bore, where some concentration of flux occurs, and this means that the radius of the flat (i.e. not bevelled) portion of the pole face must be sufficiently large compared with the radius of the bore. In general, saturation causes less difficulty in lenses for scanning electron microscopes than in those for transmission microscopes, because, as we shall see later, the former normally have longer focal lengths and therefore operate at lower flux densities.

3.2.2. The general properties of iron-cored magnetic lenses

In a lens of the type shown in fig. 3.16(*b*), the flux density at points along the axis reaches a high value in the vicinity of the air gap and is very small elsewhere. It follows that the focusing properties are determined almost completely by the form of the field in and near the gap. The region through which electrons pass is a region of free space, where Laplace's equation for the magnetic potential ψ must hold. Moreover, the system has axial symmetry and, in this case, it can readily be shown that Laplace's equation allows the magnetic potential ψ_{rz} at any point to be expressed in terms of the magnetic potential ψ_{0z} at points on the axis, by the infinite series

$$\psi_{rz} = \psi_{0z} - \frac{r^2}{2^2}\psi_{0z}'' + \frac{r^4}{2^2 4^2}\psi_{0z}'''' - \dots, \qquad (3.16)$$

where dashes indicate differentiation with respect to z. Thus, if the magnetic potential (or the axial component of flux density, from which the potential can be deduced) is known for all points on the axis, complete information about the field is available and, in principle, any electron trajectory through the lens can be determined.

In practice, mathematical difficulties are considerable and we shall not here be concerned with the techniques of computation that have been employed. Early workers used empirical equations to represent the axial flux density and, in particular, an equation first suggested by Glaser, has been widely used. In this, it is assumed that
$$B(z) = B_{M}/[1 + (z^2/a^2)], \qquad (3.17)$$

where B_M is the maximum value of the axial flux density and a is a constant depending on the geometry of the lens.

With the gradual improvement of electron-optical equipment, it became clear that the Glaser equation was not sufficiently accurate for design purposes and various attempts were made to determine the axial field by direct measurement. Finally, a comprehensive series of measurements was made by Liebmann (1955a, 1955b, 1955c) and by Liebmann and Grad (1951), using a very accurate resistance-network analogue to solve Laplace's equation. It is from these papers that we shall quote. An analogous summary of magnetic lens properties has been given by Fert and Durandeau (1967).

Since the variety of shapes and sizes of magnetic lenses is infinite and each can be used under a wide range of operating conditions, it is possible to express results in a concise form only if certain assumptions and conventions are adopted. These are

(a) The results apply only to lenses with pole pieces of the forms shown in fig. 3.17(a) (symmetrical) and fig. 3.17(b) (asymmetrical) respectively, where the bores are cylindrical and the faces FF are plane.

The rather small number of investigations on lenses with pole faces of other (i.e. non-planar) forms suggests that mild shaping of these surfaces has little effect on the properties and makes the lens more difficult to construct with the required accuracy. Plane faces in the vicinity of the bore with possible bevelling at the edges to prevent saturation have commonly been used.

(b) The flux density throughout the iron circuit is low enough to avoid saturation; this condition is satisfied in all lenses with which

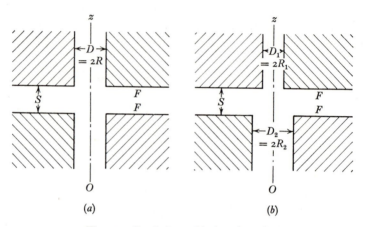

Fig. 3.17. Symbols used in lens formulae.

we shall be concerned. So long as the flux density does not exceed 1.25 Wb/m², the relative permeability is likely to be greater than 1000 and almost the whole of the magnetomotive force supplied by the magnetizing coil will appear across the air gap. If this coil has N turns carrying current I amperes, the flux density between the parallel faces of the pole pieces, at a point well away from the edge of the bore, will be given approximately by

$$B = \mu_0 NI/S = 4\pi \times 10^{-7} NI/S \, \text{Wb/m}^2, \qquad (3.18)$$

where S is measured in metres.

The maximum flux density along the axis of the lens will be somewhat less than the value given by (3.18) and, for the lens of fig. 3.17(a) will depend on the ratio S/D.

(c) To allow for effects resulting from relativity, the energy of the electrons passing through the lens is specified by a corrected voltage V_r where

$$V_r = V(1 + 10^{-6}V), \qquad (3.19)$$

and V is the actual voltage through which the electrons have been accelerated.

(d) For the symmetrical lens of fig. 3.17(a), the form of the magnetic field depends only on the ratio S/D, so long as saturation of the iron is avoided. Once the value of S/D has been chosen, the magnitude of the field can be specified in terms of either S or D.

From the elementary laws of electron dynamics it can readily be shown that, for a given value of S/D, the shape of the trajectory of an electron passing through the lens depends only on the ratio $V_r/(NI)^2$. Once again, the scale of the trajectory will be proportional to S or to D.

From the above considerations it follows that, when S/D has been chosen, any linear property of the lens, such as its focal length f, for all values of V_r, N and I, and for lenses of all sizes, can be represented by a single curve in which f/D is plotted against $V_r/(NI)^2$.

Liebmann and Grad (1951) give a number of curves of this kind from which not only the relative focal length f/D, but also the relative coefficient of spherical aberration C_S/D (§ 3.2.3), the relative coefficient of chromatic aberration C_c/D (§ 3.2.4) and other quantities can be determined for values of S/D from 0.2 to 2.0.

(In this early paper, Liebmann and Grad use as one of the variables the 'excitation parameter'

$$k^2 = \beta(NI)^2/V_r, \qquad (3.20)$$

where β is a function of S/D which can be obtained from their fig. 2. In later work the more convenient variable $(NI)^2/V_r$ or $V_r/(NI)^2$ is used.)

(e) When one turns to asymmetrical lenses, fig. 3.17(b), the rather surprising result emerges that the curves for symmetrical lenses can still be used with quite good accuracy if one substitutes $(R_1 + R_2)$ for D, where R_1 and R_2 are respectively, the radii of the two bores. Thus it is possible to plot unified curves for $f/(S + R_1 + R_2)$, $C_S/(S + R_1 + R_2)$ and so forth, which are approximately valid for both symmetrical and asymmetrical lenses (Lieb-

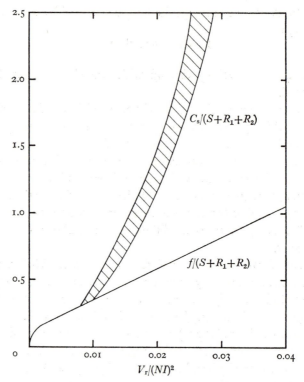

Fig. 3.18. Properties of a magnetic lens, replotted from data given by Liebmann (1955 c).

mann, 1955 c). Curves of this sort, replotted from Liebmann's data, are shown in fig. 3.18. It will be noted that, while a single curve suffices to represent the focal lengths of lenses commonly used, the spherical aberration coefficient C_s is affected to an appreciable extent by the geometrical form of the lens. Nevertheless it is still possible to include values of C_s for many commonly-used lenses in the area between two curves, shown shaded in fig. 3.18. This band is quite narrow and, for any value of $V_r/(NI)^2$, the greatest and least values of C_s do not differ by more than about ten per cent from the mean.

There is one important difference between a symmetrical and an asymmetrical lens, even when S is the same for both and when D for the one is equal to $(R_1 + R_2)$ for the other. Although f will be very nearly the same for both, this focal length is measured from the electrical centre of the lens (assuming the lens to be sufficiently weak for the principal planes to be considered coincident with the electrical centre). In the symmetrical case, the electron-optical centre coincides with the geometrical centre, but in the asymmetrical lens it does not. Thus, although f is the same for the two lenses, the position of the focus is not.

It will be appreciated that the account of Liebmann's work which has just been given represents a rather brief summary of a complex situation; complications which are not of importance in scanning electron microscopes have not been considered. In particular, no mention has been made of factors which become important when strong excitation is used and $V_r/(NI)^2$ is less than 0.009, as may happen in the objective lens of a transmission electron microscope. In such cases it is no longer sufficiently accurate to assume that the principal planes of the lens coincide with its electrical centre. Moreover, electrons from a distant source are brought to a focus at a point within the magnetic field of the lens and a more precise definition of focal length is then needed: in an asymmetrical lens, the focal length depends on the direction in which the electrons are travelling.

For our present purpose, the curves of fig. 3.18 enable a designer to select a lens geometry which will be satisfactory for the early demagnifying stages of a scanning microscope. We shall later (§ 3.2.6) consider the design of the final lens, to which special restrictions apply but, before doing this it is convenient to discuss the aberrations from which all lenses suffer.

3.2.3. Spherical aberration

In publications on electron lenses there has sometimes been confusion as to the meaning of the spherical aberration coefficient C_s (Archard, 1958), so we begin by defining this quantity.

In fig. 3.19 (a) let a point object P give rise to a Gaussian image Q. For simplicity we suppose the lens to be 'thin' so that the principal planes coincide with the plane of the lens L. Although rays making quite large angles with the axis have been drawn for the sake of clarity, it is to be understood that, in practice, none of these angles is likely to be much greater than 0.05 radian, so the usual approximations can be made.

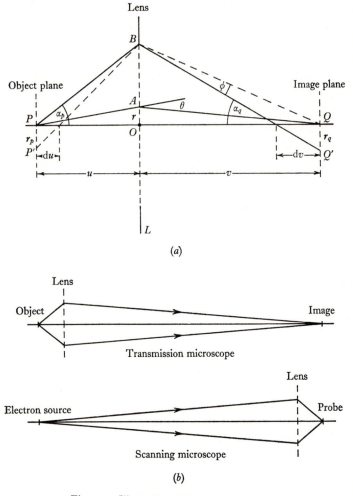

(a)

(b)

Fig. 3.19. Illustration of the definition of C_s.

Consider a ray PAQ, which intersects the plane of the lens at a very small distance r from the axis. The inclination of the incident ray to the axis is r/u and that of the refracted ray r/v. The total deviation θ of the ray resulting from its passage through the lens is $r[(1/u)+(1/v)]$. If the lens were perfect, the deviation of all such rays would be proportional to r and, since u is a fixed quantity, v would be the same for all rays. The point object P would give rise to the point image Q. In practice, the lens is not perfect, but we may write

$$\theta = a_1 r + a_3 r^3 + a_5 r^5 \dots, \tag{3.21}$$

where

$$a_1 = (1/u) + (1/v),$$

and the other constants $a_3, a_5 \dots$ depend on the values of u and v as well as on the properties of the lens. Even terms are omitted from the series because the magnitude of θ must be the same for equal positive and negative values of r. Since, in an electron lens, r is limited to values which are small compared with u and v, the terms in (3.21) rapidly decrease in value and, for most purposes, it is sufficient to include only the third-order correction $a_3 r^3$.

Consider now the ray PB, which makes the maximum angle α_p with the axis that is permitted by the aperture of the lens. If only the first term in (3.21) were operative, the refracted ray BQ would pass through the Gaussian image Q. Actually, the third-order term causes the refracted ray to intersect the plane of the Gaussian image in Q'. Thus, taking all rays starting from P, the image is spread over a circular area of radius $r_q = QQ'$.

The angle ϕ between BQ and BQ' results from the third-order term in (3.21). It is thus proportional to OB^3 and therefore to α_p^3. Again, r_q is proportional to ϕ and therefore to α_p^3. Finally, both r_q and the transverse magnification M are proportional to the image distance v. We therefore find it convenient to write

$$r_q = C_s M \alpha_p^3, \tag{3.22}$$

where C_s is a quantity having the dimensions of length, which we define to be the coefficient of spherical aberration. It is important to realize that this procedure does not eliminate the effect of the object distance u. It is true that both M and ϕ increase as u decreases, but there is no simple relation between the laws governing these two changes. Hence C_s is not a constant of the lens; it depends also on the positions of object and image. The definition of C_s is a matter of convention which, as we shall shortly see, has practical convenience.

Another useful expression involving C_s can be obtained by supposing object and image to be interchanged in fig. 3.19(a). Then a point object at Q would give rise to a circular disc image, of radius r_p in the Gaussian image plane at P. With the notation of the figure we have

$$1/v - 1/u = 1/f$$

and $$dv/v^2 = du/u^2,$$

so that $$\frac{r_p}{r_q} = \frac{\alpha_p \, du}{\alpha_q \, dv} = \frac{\alpha_p}{\alpha_q} \cdot \frac{u^2}{v^2} = \frac{v}{u} \cdot \frac{u^2}{v^2} = \frac{1}{M}. \tag{3.23}$$

We may therefore write $$r_p = C_s \alpha_p^3, \tag{3.24}$$

where all quantities now relate to conditions at P, M has disappeared from the equation and C_s depends on the distance of P from the lens. Reverting to our original arrangement with the object at P we see that, because the ray paths are independent of the direction in which they are traversed, the actual disc of confusion of radius r_q at the image, is equivalent to a hypothetical disc of confusion of radius r_p at the object. So far as the object is concerned, r_p is the quantity which will limit resolution and, as (3.24) shows, it is independent of magnification except in so far as M determines u.

As a rather striking illustration of the extent to which the coefficient of spherical aberration depends on the positions of the conjugate points, let us express r_q in terms of quantities relating to the plane at Q. By arguments similar to those already used, we have

$$r_q = C_s' \alpha_q^3, \tag{3.25}$$

where C_s' is the new coefficient appropriate to this plane. Since $\alpha_p = M\alpha_q$, comparison of (3.24) and (3.25) shows that

$$C_s' = M^4 C_s. \tag{3.26}$$

If, then, the coefficient of spherical aberration is so greatly affected by the distance of the object from the lens, how can we usefully tabulate the values of C_s for a particular lens? Fortunately the situation is saved by the fact that an electron lens is normally used only in situations where it produces either very high magnification (as in the transmission microscope) or very low magnification (as in the scanning microscope). The two situations are shown in fig. 3.19(b), where the directions of motion of the electrons are indicated by arrows. In each case we are interested in conditions in a plane which is very close to the focus; in the transmission microscope, the radius of the circle of confusion in this plane limits resolution in terms of object distances while, in the scanning microscope, it controls the effective size of the probe spot. Thus it is the value of

C_s for a plane at or near the focus of any lens that is commonly quoted and it is this value which is plotted in fig. 3.18. The radius of the circle of confusion in this plane is then given by $C_s\alpha^3$, where α is the angle which the limiting rays make with the axis, in this plane. It should always be remembered, however, that the effective coefficient of spherical aberration is likely to be appreciably greater than the normally quoted value whenever an electron lens is used under conditions where object and image distances are more nearly equal.

In the above discussion we have confined our attention to the radii of the circles of confusion in the Gaussian object and image planes respectively. However, in the scanning microscope, what is needed is the smallest possible probe spot and it turns out that the focused probe does not have its smallest radius in the plane of the Gaussian image. The reason for this is clear from fig. 3.20(a), where several rays, converging towards the image at various angles with the axis, have been drawn. Q is the Gaussian image and we wish to find the minimum radius of the beam. We have to consider rays approaching the image plane with inclinations to the axis of all values up to some maximum angle α which is determined by the aperture of the lens and the distance of the image from the lens. Considering only rays above the axis, fig. 3.20(a) shows that the minimum beam radius will be located at some point where the intersection K of (a) a ray from the lower half of the lens making angle α with the axis and (b) a ray from the upper half of the lens making some smaller angle θ with the axis, is at its maximum distance from the axis.

To determine the value of θ which satisfies this condition we take rectangular axes through Q as in fig. 3.20(b); x along the axis of the lens and y in the plane of the Gaussian image. The equations of the two rays are then

$$y = \alpha x + C_s\alpha^3$$

and

$$y = -\theta x - C_s\theta^3,$$

whence, for the point of intersection

$$x_1 = -C_s(\alpha^2 - \alpha\theta + \theta^2),$$

and

$$y_1 = C_s(\alpha^2\theta - \alpha\theta^2).$$

Differentiating with respect to θ, we find that y_1 has its maximum value when

$$\theta = \tfrac{1}{2}\alpha$$

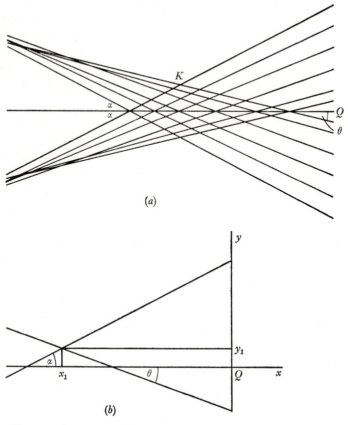

Fig. 3.20. Construction for determining the minimum probe diameter.

and
$$y_{1,\,\mathrm{max}} = \tfrac{1}{4}C_s\alpha^3. \tag{3.27}$$

$y_{1,\,\mathrm{max}}$ is the smallest radius of the beam which comes from a point object and it is this value which is used in (2.11).

3.2.4. Chromatic aberration

The focal length of a magnetic lens is a function of the maximum flux density B_0 on the axis of the lens and of the relativistically-corrected voltage V_r through which the incident electrons have been accelerated. Any variations δB_0 and δV_r in these quantities will result in changes δf in the focal length f and corresponding radial displacements δr of the point in which a ray intersects the Guassian image plane. If the ray is inclined to the axis at an angle α, we have

$$\delta r = \alpha \, \delta v, \tag{3.28}$$

where δv is the change in image distance caused by the change δf. Thus, when the variations occur randomly with time, the image of a point object will effectively extend over a circular disc of radius δr from this cause alone.

When we come to define a coefficient of chromatic aberration C_c, we meet complications analogous to those discussed in connection with spherical aberration and, as in that case, we limit ourselves to the condition that the image or object is very close to a focus. In particular, when we are concerned with the formation of an electron probe, electrons from a remote point source will be spread over a circular disc of confusion in a plane very close to the focus and (3.28) becomes

$$\delta r = \alpha \, \delta f. \tag{3.29}$$

From fig. 3.18 we see that, for any given lens with medium or weak excitation, it is a rough approximation to write

$$f = \mathrm{const}. \, V_r / (NI)^2, \tag{3.30}$$

and, if saturation is negligible, I is proportional to B_0. Hence

$$\delta f / f = (\delta V_r / V_r) - 2(\delta B_0 / B_0). \tag{3.31}$$

Combining this with (3.29), we have

$$\delta r = \left(\frac{\delta V_r}{V_r} - \frac{2\delta B_0}{B_0} \right) f\alpha. \tag{3.32}$$

Since (3.30) is only a rough approximation, we modify (3.32) to

$$\delta r = \left(\frac{\delta V_r}{V_r} - \frac{2\delta B_0}{B_0} \right) C_c \alpha, \tag{3.33}$$

where the chromatic aberration coefficient C_c will depend on the excitation of the lens as well as on its geometry. For the conditions of operation of a scanning microscope, it is sufficiently accurate to consider C_c and f to be equal but, with the stronger excitation used in a transmission microscope, the ratio C_c/f may fall to about 0.7.

Variations in both B_0 and V_r may result from imperfect stabilization of power supplies. Such variations are likely to be independent of each other and if neither is to limit the performance of the micro-scope, we must make both $\delta V_r/V_r$ and $\delta I/I$ small in comparison with $r/f\alpha$, where r is the radius of the Gaussian image of the probe. Taking as typical values for good resolution

$$r = 5 \times 10^{-9}\,\mathrm{m}, \quad f = 10^{-2}\,\mathrm{m}, \quad \alpha = 10^{-2}\,\mathrm{radian},$$

it appears that both voltage and current supplies for the final lens should be stabilized to about one part in 10^5. This is not particularly difficult to achieve over short periods, but steady drift can be more troublesome. Since the exposure time for a micrograph may be several minutes, a better specification for power supplies is to say that neither voltage nor current should vary by much more than one part in 10^6 per minute.

Assuming variations in both voltage and current to have been made negligible, we still have an effective variation δV_r resulting from the spread of the initial velocities of electrons leaving the cathode and, for a given cathode there is little we can do to reduce this. If δV_r represents the total effective spread, and if we assume the microscope to have been correctly focused for electrons of mean velocity, we may write with sufficiently good approximation, from (3.33),

$$\delta r = C_c \alpha \delta V_r / 2V_r \qquad (3.34)$$

and, for the diameter d_c of the disc of confusion

$$d_c = C_c \alpha \delta V_r / V_r. \qquad (3.35)$$

This is the expression used in (2.12).

3.2.5. Astigmatism and stigmators

In the foregoing discussion of magnetic lenses it has been assumed that the structure has perfect symmetry about its axis. In practice this will not be achieved and astigmatism will then be present. As a result, electrons from a distant point source will not be brought to a point focus, even if there is no spherical or chromatic aber-ration. If we assume the system to have elliptical rather than circular symmetry, the electrons will converge to two separate line foci, at right angles to each other and to the axis. The foci will be separated by an axial distance z_a and between them there will be a disc of minimum confusion of radius $z_a\alpha$, where α is the angle which the converging rays make with the axis. It is clearly desirable that $z_a\alpha$ should be smaller than the radius of the disc of least confusion caused by spherical aberration, $\frac{1}{4}C_s\alpha^3$.

An important source of astigmatism, which has been treated by Sturrock (1951) and Archard (1953), results from imperfect machining of the pole pieces and various defects may be present. They find that, if the axis of the bore in one pole piece is parallel to, but displaced from, the axis of the bore in the other pole piece, no harm will result so long as the displacement does not exceed 30 μm. Similarly, if the two axes are inclined to each other, the inclination should not exceed 0.005 radian. Both of these tolerances are well within the limits of normal workshop practice.

The requirements for roundness of bore and planeness of pole face are a good deal more stringent. Archard's curves show that, under the conditions likely to obtain in a scanning microscope, the distance z_a between the focal lines will be about twice the difference between the maximum and minimum radii of the bore. Similarly, z_a will be about twice the maximum distance by which the surface of a pole piece departs from the mean plane. Taking as typical values for good resolution

$$\alpha = 8 \times 10^{-3}\,\text{radian}, \quad \tfrac{1}{4}C_s\alpha^3 = 2 \times 10^{-3}\,\mu\text{m},$$

we have
$$8 \times 10^{-3}z_a < 2 \times 10^{-3}\,\mu\text{m},$$

so that the tolerances on both the radius of the bore and on the planeness of the pole faces should not greatly exceed 0.1 μm. Such accuracy is not easy to attain.

Quite apart from the difficulty of machining the pole pieces with the required accuracy, lens defects may result from inhomogeneity in the magnetic properties of the iron, from asymmetry in the winding or from charging of any contamination that may be deposited on the aperture surround. Thus it is common practice to make provision to correct defects of these kinds by the inclusion of a device known as a stigmator.

It is usually sufficient to assume that the field caused by the defects just mentioned has two-fold symmetry about the axis of the lens. Asymmetry of higher order can, of course, occur, but it is rarely troublesome. With the assumption of two-fold symmetry in the main field of the lens, the stigmator must provide a weak correcting field which also has two-fold symmetry. Either an electrostatic or a magnetic field may be used, but we shall confine our attention to electrostatic stigmators.

Since the magnitude and the direction of the asymmetry of the main field are unknown, both of these quantities must be variable in the correcting field supplied by the stigmator. A field of the appropriate form is provided by a quadrupole lens consisting of four short cylindrical rods, with axes parallel to the axis of the main lens,

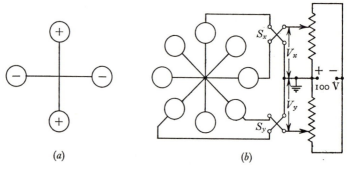

(a)   (b)

Fig. 3.21. An electrostatic stigmator. After Mulvey (1959).

fig. 3.21 (a). Opposite pairs of rods are maintained, respectively, at positive and negative potentials with respect to earth and the magnitudes of these potentials determine the strength of the correcting field. If, in addition, provision is made for the stigmator to be rotated about its axis, the direction of asymmetry of the correcting field can also be adjusted.

In practice, it is not usually convenient to arrange for the mechanical rotation of the stigmator and this is avoided in a scheme described by Mulvey (1959) and illustrated in fig. 3.21 (b). There are now eight rods, connected in pairs to variable potentials, as shown. The arrangement may be looked upon as a combination of two quadrupoles which produce astigmatic components A_x and A_y at right angles to each other. The magnitudes of these components are proportional to the voltages V_x and V_y, which can be varied and can also be reversed by means of the switches S_x and S_y. Thus, the resultant of the two components is continuously variable in both magnitude and direction.

To operate the stigmator, V_x is set to zero and V_y is adjusted to correct astigmatism as far as possible. V_x is then raised to improve the correction still further and a final adjustment to V_y may be made if necessary.

An alternative arrangement has been described by Picard (1954).

3.2.6. Design of the final lens

When the source of electrons in a scanning microscope is a thermionic cathode, two or three stages of demagnification will normally be needed to give a sufficiently fine final probe. The information already given (§ 3.2.2) suffices for the design of magnetic lenses for the early stages. Aberrations in these early lenses are rarely of importance, since the final lens accepts from them only an extremely narrow cone of electron rays. The only important electron-optical parameter is the focal length and there is wide latitude in the choice of the geometrical form of the lens. When we come to the final lens the case is quite different; aberrations are of the first importance and there are certain restrictions which do not apply to earlier stages. These we now consider.

For most purposes the operation of a scanning microscope depends on the passage of secondary electrons from the specimen to a collector which may take various forms, but which will normally be too large to mount within the bore of a lens. Secondary electrons emitted over a wide solid angle must be able to reach the collector, so it is the usual practice to mount both specimen and collector in a separate specimen chamber which is attached to the face of the lens. The aberrations of a lens increase rapidly with focal length, so the latter should be kept as short as possible. Taken together, these considerations suggest that, for the best resolution, the disposition of the components should be roughly as shown in fig. 3.22(a), where the specimen is mounted immediately outside the bore of the lens. In particular cases there may be good reasons for varying this arrangement and mounting the specimen at a greater distance from the lens. For example, one may wish to scan a relatively large area of specimen at low magnification, or to manipulate the specimen while it is under observation. The specimen chamber can be made as large as is necessary to accommodate special requirements of this kind, but it should always be borne in mind that increasing the distance between specimen and lens has an adverse effect on aberrations and therefore on the resolution of the instrument.

We shall see later (§4.5.5) that most of the secondary electrons are emitted with energies of only a few electron volts. If their passage to the collector is not to be hindered, it is therefore essential that the specimen should be mounted in a region of low magnetic field strength. It is usual to maintain the collector at a positive potential of a few hundred volts with respect to the specimen and the strength of the magnetic field that can be tolerated will depend on this voltage

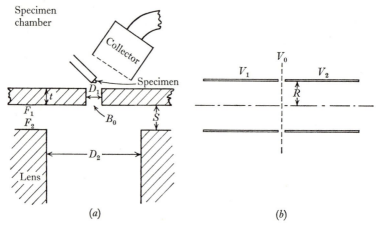

Fig. 3.22. (a) Disposition of components near the final lens. (b) Symbols used in (3.36).

and on the exact disposition of the components. A rough order-of-magnitude calculation suggests that the flux density should not greatly exceed 0.001 Wb/m². The axial flux density B_0 at a point level with the inside face F_1 of the pole piece is likely to be of the order of 0.1 Wb/m², so the thickness t of this pole piece and the diameter $D_1 = 2R_1$ of its bore must be chosen to provide an attenuation by a factor of about 100 in the flux density between the inside and outside faces.

To obtain a solution of this problem we make use of results given by Gray (1939) for the corresponding electrostatic case of the variation of potential along the axis of two adjacent equal hollow cylinders maintained at potentials V_1 and V_2 respectively, as in fig. 3.22 (b). By symmetry, the plane midway between the two cylinders is an equipotential $V_0 = \frac{1}{2}(V_1 + V_2)$. If distance z is measured from this plane, Gray finds that the potential V along the axis of the right-hand cylinder is given to a close approximation by

$$V = V_0 + (V_2 - V_0)\tanh \omega z, \qquad (3.36)$$

where $\omega = 1.32/R$. Thus the axial field E is

$$E = -\frac{dV}{dz} = -(V_2 - V_0)\omega \operatorname{sech}^2 \omega z, \qquad (3.37)$$

or, if E_0 is the field when $z = 0$

$$E = E_0 \operatorname{sech}^2 \omega z. \qquad (3.38)$$

Turning now to the magnetic lens of fig. 3.22(a), we are no longer dealing with cylindrical bores of equal radii. It will shortly appear that D_1 must be made very small in comparison with D_2, so the plane of the inner face F_1 of the outer pole piece will be very nearly a magnetic equipotential and we have postulated that the flux density in this plane will be of the order of 0.1 Wb/m². Thus for the axial flux density at points in the bore distant z from the plane of F_1, we may write

$$B = B_0 \operatorname{sech}^2 \omega z. \qquad (3.39)$$

If the flux density at the far end of the bore is to be less than $0.01B_0$, ωt must exceed 3, or $t > 1.14D_1$. Bearing in mind that the specimen will be a short distance away from the end of the bore, where the magnetic field will be somewhat weaker, we take as a convenient rule that t should not be less than D_1. For the reasons already given, the focal length should be kept as short as possible, so t also should be small. It has been common practice for both t and D_1 to have values in the region of 0.4 cm. One limitation on t is that magnetic saturation of the iron must be avoided. However, the flux to be carried increases with distance from the axis, so saturation can usually be avoided by arranging for t also to increase with this distance.

When we turn to the other pole piece of the lens, we find that the diameter D_2 of this must be made much larger than D_1. This bore may have to accommodate scanning coils and a stigmator; it will, in any case, have to avoid obstruction to pencils of electrons which have been deflected a considerable distance from the axis. For this pole piece, the bore diameter D_2 is commonly in the region of 4 cm. It is now clear that the lens will be highly asymmetrical and, in such lenses, the axial flux density reaches its maximum value quite close to the face F_1 of the pole piece containing the smaller bore. It is in this region, therefore, that the limiting aperture should be placed. Its exact position may depend on mechanical considerations.

From what has been said above, it will be clear that the ratio D_2/D_1 of the bore diameters in the final lens is quite outside the range covered by the curves of fig. 3.18. Liebmann (1955b) has studied lenses of this type, which he terms pinhole lenses, by considering the structure to be a compound of two separate lenses. The first of these includes the pole piece with large bore D_2 and extends up to the inner face F_1 of the other pole piece. It is assumed that the presence of the small bore D_1 does not greatly perturb the field in this first lens, which can therefore be considered to be one half of a symmetrical lens of bore D_2 and axial gap $2S$. This lens has focal length f_0 and spherical aberration coefficient C_{s0}. After leaving this main lens, the electrons enter the pinhole bore D_1, where the field rapidly falls to a small value, so that formulae appropriate to a thin lens can be applied. The total effect of the whole lens is then determined by adding the refractions resulting from the two component parts.

Liebmann concluded that the second lens made very little difference to the total focal length f, which was therefore very nearly equal to f_0. It did, however, increase the total spherical aberration coefficient C_s, the difference between C_s and C_{s0} being of little importance for values of $V_r/(NI)^2$ up to about 0.015, but increasing fairly rapidly for larger values. Some of Liebmann's results have been re-plotted in fig. 3.23, in which l is the distance of the focus from the inner pole face F_1. This quantity is of importance since $(l-t)$ is the working distance of the specimen from the outer surface of the lens. From these curves, certain general conclusions can be drawn for the type of lens likely to be used in a scanning microscope.

(a) The quantity $(S + R_1 + R_2)$ occurs in the denominator of each of the quantities plotted, so changes in S, R_1 and R_2, which leave their sum unaltered, will not influence the properties of the lens. Moreover, the sum will usually be dominated by R_2, so changes in S or R_1 will have little effect on C_{s0}, f_0 or l.

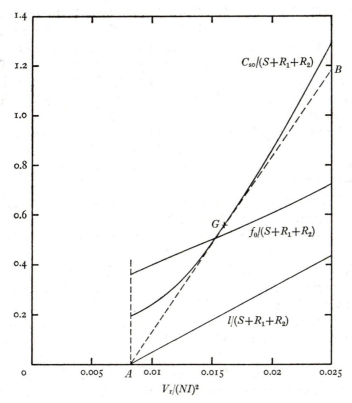

Fig. 3.23. Properties of a pinhole lens, replotted from data given by
Liebmann (1955*b*).

(*b*) For a given lens, as the quantity $V_r/(NI)^2$ is reduced, both
C_{s0} and l decrease. Thus, it is always advantageous to operate
with l, and therefore the working distance $(l-t)$, as small as
possible.

(*c*) In general, we wish to make the ratio l/C_{s0} as great as possible
and this imposes a limitation on the value of $V_r/(NI)^2$. In fig. 3.23
the curve for $l/(S+R_1+R_2)$ is a straight line. From the point A,
where this line cuts the axis of $V_r/(NI)^2$ a straight line AB has been
drawn to be tangential to the curve for $C_{s0}/(S+R_1+R_2)$ at the
point G, corresponding to a value of 0.016 for $V_r/(NI)^2$. This value
clearly makes l/C_{s0} a maximum.

Accepting the above conclusions, the design of a lens might proceed somewhat as follows. The minimum value of the working

distance $(l-t)$ will depend on the specimen that we wish to examine and will be included in the initial specification of the lens: we assume it to be 0.3 cm. Similarly, t must be large enough to prevent magnetic saturation of the iron in the pole piece and a value of 0.3 cm will be assumed for this also. We thus arrive at 0.6 cm for l. If $V_r/(NI)^2$ is to have its optimum value of 0.016, $l/(S+R_1+R_2)$ will be equal to 0.21, giving values of 2.9 cm for $(S+R_1+R_2)$ and 1.6 cm for C_{s0} respectively. We have seen that, to keep the magnetic field at the specimen to an acceptably low value, R_1 should not exceed $\frac{1}{2}t$, so we put $R_1 = 0.15$ cm, leaving $(S+R_2)$ equal to 2.75 cm. The division of this sum between S and R_1 can be varied to suit mechanical considerations, but $S = 0.75$ cm and $R_2 = 2.0$ cm would meet many requirements. Finally, from fig. 3.23, f_0 will be equal to 1.5 cm and the coefficient of chromatic aberration C_c will not differ significantly from this value (§ 3.2.4).

A lens constructed to the above design will give good results, but there may be appreciable errors in the values deduced for C_s and f. Some spherical aberration will be produced by the passage of the electron rays through the thin lens resulting from flux in the pinhole, and this has not been included in C_{s0}. Liebmann's paper indicates that the total C_s might exceed C_{s0} by something of the order of twenty per cent when $V_r/(NI)^2$ is equal to 0.016. We shall not pursue this matter, since doubts have been cast on the validity of Liebmann's treatment of the correction. Quite apart from this, Liebmann's results make no pretence to high accuracy; their value lies in giving us a comprehensive picture of the way in which variation of any of the parameters is likely to affect the overall properties of a lens. Computer programs are now available by means of which the properties of any given lens can be determined accurately and rapidly. Thus, when a first design has been prepared from Liebmann's curves, values of f, l and C_s can be checked. Similarly, the possibility of improving the design by altering any of the parameters, can be investigated. A number of computations of this kind, Munro (1971), indicate that the procedure outlined above leads to a first design which is not very far from optimal.

From what has been written above, it will be clear that there is no single 'best' lens for all conditions that may be encountered in a scanning microscope. With any lens, spherical aberration increases with working distance, but the relation between these two quantities depends a good deal on lens geometry. In general, the geometry which yields lowest spherical aberration for short working distance will not be the same as that which gives best results when the working distance is much longer. Some experimental data on lenses, at excitations corresponding to long working distance, have been published by Barnes and Openshaw (1968).

3.2.7. Apertures

To reduce spherical aberration in the final lens to an acceptable value, the aperture of this lens must be limited by the insertion of a metal disc with a circular hole. Our earlier discussion (§ 2.6) has shown that, for any given conditions of operation, there is an optimum value α_{opt} for the semi-angle of the cone of electrons converging on the specimen. For work at the highest resolution, this angle is likely to lie in the region of five milliradian so that, assuming a focal length of 1.0 cm, the diameter of the aperture hole should be about 100 μm. It is important to note that, when longer working distances, and therefore longer focal lengths are needed, the optimum diameter of the aperture hole will change. We have already seen that C_s gets larger as the working distance increases, so α_{opt} decreases. Furthermore, for a given value of α_{opt}, the aperture diameter is proportional to the focal length and therefore approximately proportional to the working distance. A further complication is that it may sometimes be desirable to use an aperture whose diameter is less than the 'optimum' value, in order to increase the depth of field.

The aperture disc is usually mounted close to the face F_1 of that pole piece of the final lens which contains the smaller bore (fig. 3.22(a)). As the above discussion has shown, aperture holes of different diameter will be needed for different conditions of operation, so it is usual to mount several discs in a tray and to provide mechanical means of bringing any one of them into operation by adjustments made to controls outside the vacuum system. If unnecessary increase in spherical aberration is to be avoided, the centre of an aperture hole must lie on the axis of the electron lens. Perhaps a reasonable tolerance would allow the centre of the hole to depart from the axis by a distance not greater than about ten per cent of the diameter of the hole. For high resolution this means that the aperture must be centred with a precision of about 10 μm.

3.2.8. Lenses without iron cores

The previous discussion of electron lenses has been confined to types in which iron pole pieces and shrouds are used to concentrate the magnetic flux. We complete our survey by mentioning briefly

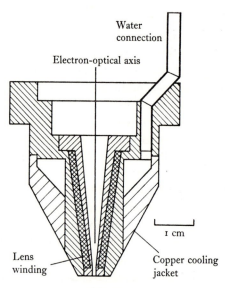

Fig. 3.24. The cross-section of a mini-lens. Redrawn by permission from Duncumb (1969).

two fairly recent developments relating to lenses which contain no ferromagnetic material.

A lens of this type has come to be known as a mini-lens, since it is very much smaller in size than a conventional iron-cored lens with similar properties. The cross-section of a mini-lens constructed by Fontijn (Duncumb, 1969) is shown in fig. 3.24: the overall diameter of this lens is only 5 cm and the large current density necessitates water-cooling of the windings. Such lenses have already found application in X-ray microanalysers, where their small size greatly simplifies the problem of detection of the X-rays (Fontijn, Bok and Kornet, 1969). At their present stage of development, astigmatism is considerably higher than with iron-cored lenses, but this is a constructional rather than a fundamental limitation. For scanning electron microscopy a more serious limitation would be the relatively strong magnetic field at the specimen. So far as is known, mini-lenses have not yet been used in scanning microscopes, but, at a later date, they may find application in special-purpose instruments.

A rather similar development has been reported by Bassett and Mulvey (1969). In this case the coil has a flat 'pancake' form and theoretical studies of its properties have been made. The results are promising and experimental work is in progress. The authors point out that lenses without iron cores should have particularly valuable properties if operated at a sufficiently low temperature for the winding to be superconductive, since much larger current densities would then be possible. Once again, the existence of a high magnetic field at the specimen is likely to make 'pancake' lenses more useful for X-ray microanalysis than for scanning microscopy.

3.2.9. Microscopes with field-emission electron sources

The preceding discussion of electron lenses has been based on the assumption that electrons are produced by a thermionic cathode and that the effective source is the crossover produced by the electrode structure of the gun. Under such conditions, two or more stages of demagnification are needed to produce an electron probe of sufficiently small diameter.

When a field-emission source is used, the conditions are quite different and it has already been pointed out (§ 3.1.6) that the effective diameter of the source is then so small that additional demagnification may be unnecessary. It thus becomes possible to design an electron gun which both accelerates the electrons and focuses them directly on to the specimen without the use of any auxiliary lenses. This has been done by Crewe, Eggenberger, Wall and Welter (1968), who give details of the electrode structure which they used.

3.3. The scanning system

3.3.1. General considerations

Provision must be made to scan the electron probe of the microscope over a rectangular raster and, in principle, the beam may be deflected either electrostatically or magnetically. In practice, magnetic deflection has significant advantages and is generally used. When an electron beam has to be deflected through relatively large angles, it is well known (Moss, 1968) that a magnetic system causes less deflection–defocusing than an electrostatic system of comparable size. This may not be important for the deflection of the electron probe in the microscope, where angles of deflection are usually very small, but it will certainly influence the choice of cathode-ray tube to be used in the display unit. In fact, nearly all high-quality cathode-ray tubes that are available commercially use magnetic deflection. Although it would not be impossible to deflect the probe electrostatically, while retaining magnetic deflection for the cathode-ray tube, a mixed system of this kind offers no advantage and it is obviously much simpler to use magnetic deflection in both cases.

We turn next to the location of the scanning system and we consider first the situation that normally exists when high resolution is desired. Our discussion of the design of the final lens (§ 3.2.6) has shown that, if the coefficient of spherical aberration is to be kept low, the working distance between the object and the external face of the lens must be small. There is therefore no room for the scanning coils in this region and the beam must be deflected before it enters the final lens.

Following McMullan, the arrangement of fig. 3.25 is commonly used, with two sets of deflecting coils, A and B. The beam is deflected twice through angles such that the ray which originally coincided with the axis of the beam, passes through the centre of the aperture of the lens, thus keeping aberrations in the lens to a minimum. To fulfil this condition the deflecting angles θ_A and θ_B must be related to the effective distances, h_A and h_B respectively, of the coils from the plane P of the aperture, by the equation

$$(h_A - h_B)\tan\theta_A = h_B\tan(\theta_B - \theta_A). \qquad (3.40)$$

If the deflections are sufficiently small for the approximation $\tan\theta = \theta$ to be valid, this becomes

$$\theta_B/\theta_A = h_A/h_B. \tag{3.41}$$

Assuming the two coils to be of the same geometrical form, the deflections which they produce will be proportional to the ampere-

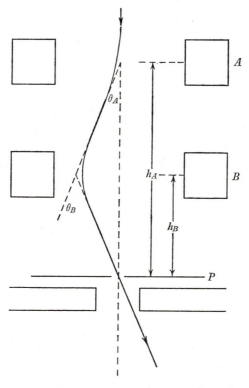

Fig. 3.25. The arrangement of the scanning coils.

turns in their windings. It is convenient to connect the coils in series so, with the same current flowing in both, the numbers of turns n_A and n_B should satisfy the relation

$$n_B/n_A = h_A/h_B. \tag{3.42}$$

This condition is independent of the magnitude of the final deflection.

For larger angles of deflection (3.41) may not be sufficiently accurate and we must satisfy (3.40). For arbitrary values of h_A and h_B the ratio n_B/n_A would then be a function of the final deflection and (3.40) could not be satisfied over the whole of the scan. However, if we make

$$h_A = 2h_B, \quad \theta_B = 2\theta_A, \quad n_B = 2n_A, \qquad (3.43)$$

we see that (3.40) is satisfied for all values of θ_A.

In practice, other considerations may have to be taken into account. If θ_A is to be greater than a few degrees, coil B must be placed fairly close to the aperture plane P, so that the maximum departure of the electron beam from the axis of the system shall not be too great. The coils themselves must have appreciable size and it may be mechanically impossible to satisfy (3.43). Fortunately, the approximation of (3.41) is nearly always good enough, so that (3.43) represents a condition which should be satisfied as far as possible, but which is not of the first importance.

The above discussion has been somewhat simplified by the assumption that the deflections occur only in or near the planes of the respective coils, whereas actually the field of each coil extends indefinitely in each direction, though with rapidly decreasing strength. The coils should be spaced apart, so that interaction between them is small, but it may nevertheless be found that a system constructed in accordance with the equations that we have derived does not quite satisfy (3.40). When necessary, a small correction can be applied by shunting one or other of the coils to reduce the current through it by an amount which can be determined by experiment. With zero deflecting current, the electron beam is adjusted to lie along the axis of the microscope, so that the maximum current passes through the aperture. The beam is next given static deflections in one plane only and the shunt is varied until maximum current once more passes through the aperture.

3.3.2. The design of the scanning coils

If V (volts) is the voltage through which the electron beam has been accelerated and b (metres) the effective length of the deflecting magnetic field through which the beam passes, it can readily be shown that the field strength B(Wb/m^2) needed to turn the beam through an angle θ (radian) is given by

$$B = 3.3 \times 10^{-6} V^{\frac{1}{2}} \theta / b. \qquad (3.44)$$

In practice B will rarely be uniform and will certainly not be strictly confined to a definite length b, but (3.44) nevertheless gives a useful indication of the field strength required. Substitution of typical values suggests that B is unlikely to be much greater than 5×10^{-3} Wb/m^2.

Air-cored coils and coils with ferrite cores have both been used successfully in scanning microscopes and we first consider the relative advantages of the two types. If the microscope is required to give low values of magnification, θ may be as large as 0.1 radian or more and, as we have already seen, it is then desirable for coil B, in fig. 3.25, to be mounted fairly close to the aperture of the final lens. It will then lie in a region where the field of the lens is strong, so no magnetic material, such as a ferrite core, can be used in its construction. To secure a reasonably uniform field, two air-cored coils approximating to a Helmholtz pair are normally used, though the coils may well have rectangular rather than circular form, in order to secure a greater length of deflecting field. Two pairs, for the two directions of deflection are needed at A and a similar set, with more turns, at B.

When low values of magnification are not essential, the coils can be placed much further away from the aperture of the lens, in a region where the magnetic field of the lens is negligible. It is then permissible to use ferrite-cored coils, which make more efficient use of the magnetic flux generated. In consequence, the inductance of a ferrite-cored coil is generally only about one third of that of the corresponding air-cored coil. This may be important if rapid scanning is required. It is probable, also, that raster distortion, defocusing and non-linearity of sweep are more serious with air-cored than with ferrite-cored coils. However, comparative experimental data on these matters are lacking for the rather small angles of deflection used in a scanning microscope. The following information on the design of ferromagnetic-cored coils is taken from a paper by Woroncow (1947).

The simplest coil of this type, for deflection in one plane only, is

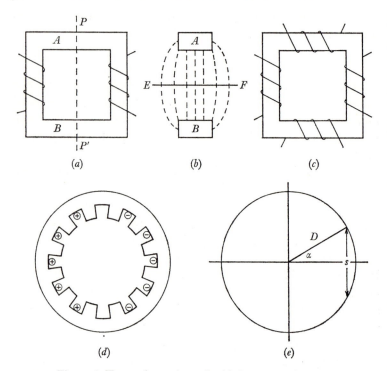

Fig. 3.26. Types of scanning coil with ferromagnetic cores.

shown in fig. 3.26(a). The core has the shape shown, in the plane of
the paper, and has rectangular cross-section at right angles to this
plane. Two opposite limbs carry equal windings connected to produce opposing magnetomotive forces so that, when a current flows
through them, a very nearly constant difference of magnetic potential is set up between the other two limbs A and B. If we take a cross-
section, at right angles to the plane of the paper, through the line
PP' the form of the magnetic field produced by this difference of
potential will be roughly as shown in fig. 3.26(b), where the axis of
the undeflected electron beam is indicated by the line EF. The field
consists of a very nearly uniform part in the space enclosed by the
core, together with a non-uniform field outside this space and it is
this non-uniform part which is chiefly responsible for any defects
which the coil may possess. Moreover, the form of the non-uniform
part of the field depends very much on the distribution of the turns

of the windings along the core and any attempt to treat the matter mathematically encounters formidable difficulties. The best form of winding must therefore be found by trial.

So far, we have considered windings to deflect the electron beam in one plane only, but it is clear that similar windings could be placed on the remaining two limbs of the core to cause deflection in a second plane, at right angles to the first, as is needed in the scanning microscope. When this is done, it is found that the best form for the windings is not necessarily quite the same as that which would be inferred from experiments in which only one pair of windings was used. Experiments to find the best design of coil should therefore be conducted with a complete set of windings, using simultaneous deflections in the two planes at right angles.

For all ordinary uses of the scanning electron microscope, coils of the type just described appear to be completely satisfactory and they are simple to construct. Woroncow states that better coils can be made by using a slotted core as in fig. 3.26(d). For a circular core, fig. 3.26(e), the length s of magnetic path in air or vacuum, corresponding to angle α is

$$s = 2D \sin \alpha. \tag{3.45}$$

Hence, for a uniform field, the total number of turns from zero to α should be proportional to $\sin \alpha$. With the slotted core of fig. 3.26(d) a three-step approximation to this law can be achieved and a field of high uniformity results. Should experimental tests reveal the presence of deflection, defocusing, distortion or non-linearity of deflection, empirical corrections can be made by slightly varying the numbers of turns in the different slots.

Anyone who intends to construct scanning coils of the types that we have described, would be well advised to consult Woroncow's paper, which contains a great deal of practical information that cannot readily be summarized.

3.3.3. Magnification

The operation of a scanning electron microscope requires exact synchronism between the deflection of the electron probe over the specimen and the deflection of the electron beam in the cathode-ray tube of the display unit. This is most readily achieved by connecting the two sets of scanning coils effectively in series and supplying them from a single generator for the X-scan and a separate single generator for the Y-scan. The outputs from these generators are determined by the magnitudes of the currents needed to produce a raster of the required size on the cathode-ray display tube; currents required to deflect the electron probe are usually a good deal smaller. It is therefore necessary to reduce the currents flowing through the scanning coils in the microscope by the insertion of suitable attenuating networks which can, when necessary, be designed to make allowance for the inductances of the coils, so that the waveforms of the deflecting currents are not distorted.

The magnification of the microscope is the ratio of the length of one side of the raster on the display cathode-ray tube to the length of the corresponding side of the raster scanned by the electron probe on specimen. It is one of the most important advantages of the scanning microscope that the magnification can rapidly be changed over a range of perhaps 20 to 5×10^4 times without refocusing and this means that provision must be made to vary the currents through the scanning coils in the microscope over equal ranges. To cover such a large range it is common practice to use an attenuating network with two controls; the first allows the magnification to be varied in steps, while the second gives continuous variation over the interval from one step to the next. The design of circuits to achieve this end presents no great difficulty and we shall not discuss it further. The calibration of such circuits in terms of magnification is, however, more complicated, because the length of the side of the raster scanned on the specimen is not determined by the current in the scanning coil alone. For a fixed amplitude of current in the coil, the length is proportional to the distance of the specimen from the aperture of the final lens and inversely proportional to the square root of the voltage through which the electrons have been accelerated.

To cater for variation of the accelerating voltage, it is common practice to provide calibration scales relating to a number of fixed values of this voltage. In the same way, the scales can relate to particular distances from object surface to lens aperture, but the adjustment of the object to the correct distance is less easy, particularly when its surface is not at right angles to the axis of the electron beam. If the object carriage permits variation of the distance from the aperture, the value of the final-lens current required for focusing can be used as an indicator of the correct position.

For many purposes the above difficulties are of little importance, since approximate values of the magnification suffice. For quantitative work a direct calibration has much to recommend it. At low magnifications a comparison can be made with the micrometer movement of the object carriage. If, then, the attenuator used to reduce the current to the scanning coils is itself calibrated, higher magnification values can be deduced.

When, as is often the case, the surface of the object is not at right angles to the axis of the electron beam, the magnifications will usually be different in the two directions of scan. Moreover, since the scan directions suffer rotation as a result of the passage of the electron beam through the lens, and since the extent of this rotation depends on the strength of the lens current (and thus on the position of the object), it is not easy to mount the object in such a way as to ensure that one of the scan directions coincides with the direction of greatest magnification. The most straightforward solution to this problem is to arrange for the directions of scan to be rotatable about the axis of the microscope (§ 3.3.5). It would, of course, be possible to achieve the same end by rotation of the specimen, but this would normally entail a corresponding rotation of the secondary-electron collector and, as we shall see later, this is unlikely to be convenient.

With the above arrangement the greatest and least magnifications can, by trial, be set to coincide with the two directions of scan on the object and hence with the X and Y deflections on the display unit. The values of these two magnifications can then be indicated on any micrograph that is taken.

It is convenient, at this point, to consider typical numerical values to get an idea of the magnitude of certain quantities involved in the scanning process. Fig. 3.27 represents the conditions in the vicinity of the specimen and, for the time being, we assume the surface of the specimen to be normal to the axis of the microscope. Suppose first that we wish to operate the instrument at high magnification, with the best possible resolution. It will be advantageous to mount the specimen as close to the pole piece as is convenient and typical values of the quantities shown in the figure might be

$$m = 0.2 \text{ cm}, \quad t = 0.3 \text{ cm}, \quad n = 0.5 \text{ cm},$$

giving $l = 1.0$ cm.

For the length of the side of the raster on the cathode-ray display tube we take 10 cm so that, for a magnification of 10^4, g must be equal to 10^{-3} cm, giving $\beta = 5 \times 10^{-4}$ radian.

Before examining the specimen in this way we should probably wish to make a more general survey of its surface at much lower magnification and fig. 3.27 shows that the maximum value of β, the angle of scan, is determined by interception of the electron beam by the pole piece. If we put R, the radius of the bore equal to 0.15 cm and use the values previously assumed for the other quantities, we

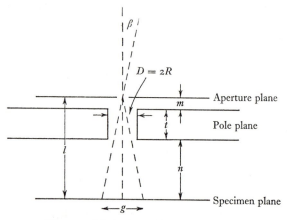

Fig. 3.27. Limitation of the angle of scan.

find $g = 0.6$ cm. This corresponds to a minimum magnification of 16.7 and a value of β of 0.3 radian. No significance is attached to the particular values that we have chosen and, in any given instrument, there may be good reasons for selecting different ones, but our discussion suffices to show the orders of magnitude of the various quantities.

3.3.4. Aberrations resulting from the scanning process

(a) Lens aberrations

Even if the scanning coils were perfect, the deflection of the beam would introduce aberrations which we must now consider. The first of these arises because the beam passes through the final lens at an angle to the axis to form an image at a distance from the axis, whereas our previous treatment of spherical aberration related to an axial beam forming an image on the axis. In other words, we are now effectively dealing with an extended object and an extended image and we must take account of all the new aberrations (distortion, astigmatism, curvature, coma, anisotropic distortion, anisotropic astigmatism and anisotropic coma) which this entails. We shall not need to analyse these terms in detail and we proceed as follows.

Let r be the distance from the axis of a point Q in the Gaussian image. In practice a ray from the corresponding point of the object will, as a result of aberrations, intersect the Gaussian image plane at some other point Q', where $QQ' = \delta r$. Let this ray approach the image at an inclination α to the axis. It can then be shown that δr can be resolved into two rectangular components δx and δy, in the image plane, such that

$$\delta x = A_1 r^3 + A_2 r^2 \alpha f + A_3 r \alpha^2 f^2 + A_4 \alpha^3 f^3, \qquad (3.46)$$

$$\delta y = A_5 r^3 + A_6 r^2 \alpha f + A_7 r \alpha^2 f^2, \qquad (3.47)$$

where f is the focal length of the lens and it is assumed that the object is at a great distance. Of these terms, $A_4 \alpha^3 f^3$ is the spherical aberration that we have already considered, but relatively little has been published about the values of the other coefficients. However, Liebmann (1955a) has shown that, for a number of lenses, A_1, A_2, A_3, A_5, A_6 and A_7 are all of the same order of magnitude as A_4. We may therefore conclude that, so long as r is small in comparison with αf, the spherical aberration will be the dominant term. Otherwise expressed, this means that spherical aberration will predominate so long as r is small in comparison with the radius of the aperture; an understandable conclusion, since the maximum additional inclination of a ray to the axis, caused by the displacement of the image point from the axis, will then be small in comparison with the maximum inclination already permitted by the aperture for an on-axis image point.

Reverting to fig. 3.27 and the numerical values that we have already assumed, we see that aberrations other than spherical may be expected to become significant if g exceeds the diameter of the aperture. For high resolution this diameter might be about 100 μm and the same value for g would correspond to a magnification of 1000. At this magnification the smallest detectable distance on the cathode-ray tube of the display unit – say 100 μm – corresponds to 0.1 μm on the specimen. This is more than ten times as great as the diameter of the disc of confusion caused by spherical aberration, so the other aberrations are unlikely to be troublesome.

Summarizing the above discussion, the situation is as follows. If we set the microscope to operate at high magnification and high resolution, the specimen will be placed at a distance of about 1 cm from the aperture of the final lens and the probe diameter will be about 0.01 μm. The diameter of the disc of least confusion caused by spherical aberration will be less than this and all other aberrations will be negligible. If we now reduce the magnification, non-spherical aberrations will increase and, when the magnification has fallen to 1000, their effect is likely to be greater than that of spherical aberration. However, by this time, we can tolerate an aberration disc of about 0.1 μm without loss of resolution, so no difficulty arises. If the magnification is reduced still further the non-spherical aberrations will predominate and, incidentally, the approximations on which (3.46) and (3.47) are based will cease to be valid. However, it is a matter of practical experience that the overall aberration usually remains below the limit at which it would impair resolution.

It should not be inferred from the above that non-spherical aberration is unimportant. At whatever magnification we wish to work, it is advantageous to use the largest aperture consistent with adequate resolution, since this gives the best signal/noise ratio or, alternatively, the shortest recording time. At high magnification, aperture size is limited by spherical aberration and the other aberrations can be neglected. At sufficiently low magnification, the reverse is likely to be true and it will then pay to increase the distance between specimen and aperture. This will increase the spherical and reduce the non-spherical aberrations to somewhat unpredictable extents, since the excitation of the final lens will have to be reduced to increase its focal length. In general terms we may expect the optimum condition to be that in which spherical and non-spherical aberrations make roughly equal contributions to the disc of confusion. Fortunately the optimum is a fairly broad one; it can be found by trial, so long as the operator has a clear appreciation of the factors involved.

(b) Depth of field

In fig. 3.28 (a), let electrons from a point source P be brought to a point image Q by the final lens of the microscope which, for the present, we assume to be perfect. If the specimen surface is to be at right angles to the electron probe, we should wish to mount the specimen so that this surface lies in the image plane at Q, but it is of interest to know by what distance the surface can be displaced from this plane without appreciably affecting the performance of the microscope. Let the planes at Q' and Q'' respectively represent the limits of displacement from this point of view. Then $Q'Q''$ is known as the depth of field of the microscope. It is important because it determines the greatest roughness of surface that can be viewed without loss of resolution at some points.

In the planes at Q' and Q'' there will be discs of confusion instead of point images and we will arbitrarily assert that the diameter of these discs should not exceed the smallest distance that we wish to resolve. Let δ be the smallest distance that can be resolved on the cathode-ray tube of the display unit and let the microscope be operated at magnification M. Then we may write

$$Q'Q\alpha = QQ''\alpha = \delta/2M,$$

or $\qquad$ depth of field $= Q'Q'' = \delta/M\alpha.$ $\qquad$ (3.48)

Taking as typical values for operation at high magnification

$$\delta = 10^{-2}\,\text{cm}, \quad \alpha = 10^{-2}\,\text{radian}, \quad M = 10^4,$$

we have $\qquad Q'Q'' = 1.0\,\mu\text{m}.$

It is one of the most valuable features of a scanning electron microscope that its depth of field is some hundreds of times as large as that obtained with an optical microscope operated at the same magnification.

So far we have considered only the focusing of a stationary axial probe: when scanning is taken into account, other complications arise. In the first place, lens aberrations will cause the different positions of the focused image to lie in a curved surface rather than a plane (fig. 3.28(b)). However, the curvature is usually sufficiently

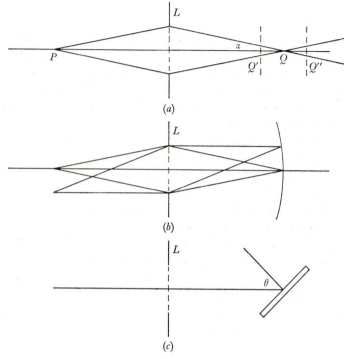

Fig. 3. 28. Aberrations resulting from scanning.

small for the focused image to lie within the limits set by the depth of field and this defect is rarely important.

Much more serious is the situation that arises when, as frequently happens, the normal to the surface of the specimen is inclined at angle θ to the axis of the microscope, as in fig. 3.28(c), to facilitate collection of secondary electrons. We assume that both θ and $(\frac{1}{2}\pi - \theta)$ are very large compared with the angle through which the beam is scanned. If D is the length of side of the raster in the display unit, D/M is the distance at right angles to the axis through which the focused probe moves. Corresponding to the extreme positions of the probe there is an axial displacement of the surface of the specimen (assumed plane and smooth) of $(D\tan\theta)/M$ and this should not exceed the depth of field. Substituting from (3.48), we have

$$\tan\theta \ngtr \delta/D\alpha. \tag{3.49}$$

Again taking, as typical values for operation at high magnification,

$$\delta = 10^{-2}\,\text{cm}, \quad D = 10\,\text{cm}, \quad \alpha = 10^{-2}\,\text{radian},$$

we find $\tan\theta \ngtr 0.1$.

This leads to too small a value of θ to permit efficient collection of secondary electrons and we can proceed in one of two ways; we can accept some degradation of resolution in the image or we can provide automatic focus-correction.

As a rule the former procedure is followed and, in practice, the degradation of resolution need not be too severe. δ can be allowed to take values two or three times as great as those which we have assumed and α can be made somewhat smaller; the image will, in any case, be properly focused in the centre. The slight defocusing at the edges is not usually sufficiently great to be troublesome.

If it is decided to correct this defect, account must be taken of the facts that the specimen is being scanned in two mutually perpendicular directions, and that neither direction will necessarily lie in the plane containing the normal to the surface and the axis of the instrument. If one of the scan directions could be adjusted to lie in this plane, the problem would be greatly simplified, since the whole of the defocusing would then result from this scan alone and the scan in the perpendicular direction could be ignored. As we have already seen, it is not easy to set up the specimen in such a way as to achieve this end, since the directions of scan depend on the excitation of the final lens and hence on the position of the specimen relative to the lens. It is therefore convenient to make provision for changing the directions of scan after the specimen has been placed in position. This could be done by arranging for the complete scanning system to be rotatable about the axis of the microscope, with a mechanical drive operated from outside the vacuum enclosure. Alternatively, and more simply, the rotation can be produced electrically, as follows.

Let x and y, respectively, be the directions of the magnetic fields produced by the two pairs of a set of scanning coils and let I_1 and I_2 respectively be the currents from the line and frame waveform generators. Ganged sine/cosine potentiometers, with suitable buffer amplifiers, are inserted between the generators and the coils in such a way that the components of magnetic field are given by

$$kH_x = I_1 \cos\theta + I_2 \sin\theta, \Big\}$$
$$kH_y = I_1 \sin\theta - I_2 \cos\theta, \Big\} \qquad (3.50)$$

where k is a constant and θ represents the angular setting of the sine/cosine potentiometers. Thus the directions of the line and frame total fields can be rotated by varying θ, while the two fields remain perpendicular to each other.

When such a system is available, the correct setting is that in which, on the display unit, the magnification is greatest in the direction of the line-scan and least in the direction of the frame-scan. The direction of the frame-scan in the microscope is then the direction of greatest slope on the specimen and it is current of the same waveform as that producing the frame-scan that must be used to correct the focal length of the final lens. The frame-scan rather than the line-scan is chosen for the correction because the frequency of the former is much lower and the inductance of the winding on the lens is correspondingly less important. For the same reason it would seem desirable to pass the correcting current through a separate winding on the lens, of relatively low inductance, rather than to attempt to vary the main lens current. The current through the correcting winding must be variable and its magnitude must be determined by experiment.

From the above discussion it will be clear that correction for this particular form of defocusing is not entirely straightforward. Relatively little work on the subject has been reported and it is too early to decide whether the results that might be obtained justify the additional complication.

3.3.5. The examination of large specimens

The foregoing discussion of scanning systems covers most ordinary uses of the microscope, in which a relatively small object is first examined at low magnification to select points of interest which, in turn, are subsequently viewed at higher magnification. Occasionally, however, one may wish to examine the whole surface of a large plane object. In such a case the object will usually be placed at an appreciable distance from the final lens to avoid an excessively large angle of scan. There will then be ample room to accommodate what-ever secondary-electron collector is used and the object can be mounted with its surface normal to the axis of the microscope.

With such an arrangement there is no reason why the scanning system should not be located between the final lens and the object. Only one set of coils would then be needed and aberrations resulting from the passage through the lens of rays making large angles with the axis, would be avoided. So far, a scheme of this kind does not appear to have been used to any great extent and details are lacking of the results that might be obtained with it. Nevertheless it seems likely, in the future, to find application to problems of the type described.

3.4. Signal current collection and amplification

3.4.1. General considerations

From our earlier discussion on the fundamental limitations on the performance of a scanning electron microscope in chapter 2, it is clear that, when the instrument is operated at high magnification, beam currents of the order of 10^{-11} A will be used. At lower magnification, larger currents can be employed with advantage. The secondary current leaving the specimen is likely to be of the same order of magnitude but, in many instances, only a fraction of this current will be collected. We are thus faced with the problem of deriving an output signal, to be passed to the display unit, from an input current which may not be much greater than 10^{-12} A.

Clearly, a high degree of amplification is needed and we must consider what general properties the amplifier should possess. Dealing first with the question of band-width, suppose that a complete picture frame of N lines is scanned in time t. The waveform of the signal entering the amplifier will depend on the way in which the collected electron current from the specimen varies, as the incident probe passes from one picture element to the next, but we may expect it to contain components of the highest frequency when successive picture elements are alternately 'black' and 'white'. In such a case, if the incident probe size were negligible, the input signal would be a square wave in which the 'on' and the 'off' portions each lasted for a time t/N^2. The period of the signal would thus be

$2t/N^2$ and the fundamental component in its frequency spectrum $N^2/2t$. In practice, the diameter of the probe is likely to be comparable with the width of a picture element on the specimen and the current density in the probe will not be constant across its area. Thus, in the case just discussed, although the signal to be amplified still has fundamental frequency $N^2/2t$, its waveform, and therefore the amplitudes of higher harmonics, will depend as much on the properties of the electron probe as on those of the specimen surface. We therefore conclude that nothing is to be gained by amplifying frequency components higher than

$$f_2 = N^2/2t, \qquad (3.51)$$

and we take this to be the upper limit of the required band-width.

Thus, a 300-line picture scanned in two seconds for visual observation would use a band-width of about 22.5 kHz. On the other hand, a 1000-line picture recorded photographically in 100 seconds, would require only 5 kHz. Excessive band-width is to be avoided, since it introduces unnecessary noise. At the lower end of the frequency range we must allow for the fact that the total time taken to scan the object may be as long as 100 seconds or more and that the average level of brightness may change from a minimum at one side of the frame to a maximum at the other. Thus the output signal from the electron collector may have a repetition frequency of 0.01 Hz or lower and will contain important components of this order of frequency, which must be passed on to the display if a correct reproduction of tonal values is to be achieved.

One method of dealing with this problem is to employ d.c. amplification throughout and this presents little difficulty when the electron detector itself provides much of the required gain. This is the case, for example, when the detector consists of a combination of scintillator and photomultiplier, or of a channel multiplier. With other detectors, however, most of the gain must be supplied by conventional amplifiers, in which a.c. coupling is a great deal more convenient than d.c. coupling. We may then employ the standard television technique of d.c. restoration or clamping to avoid the necessity of amplifying frequencies as low as 0.01 Hz, which would present considerable difficulty. In this technique, the low-frequency cut-off of the a.c. amplifier is chosen to be sufficiently low to cater for the lowest frequency occurring during the scan of a single line of the raster. This frequency will be N times as great as the lowest frequency occurring during the scan of a complete frame, where N is the number of lines in the raster. It is therefore unlikely to be less than about 10 Hz and the design of an a.c. amplifier with a low-frequency cut-off in this region is fairly straightforward. Such an amplifier will faithfully reproduce the variations in intensity during a single line, but will not take account of the much slower variations in average intensity as the complete raster is scanned. This latter problem is overcome by arranging that, during the fly-back between each line and the next, the input signal is effectively reduced to zero and the output of the amplifier is automatically set to a fixed value corresponding to 'black' on the cathode-ray tube display. In this way, any drift in black level from one line to the next is cancelled. The way in which this technique operates is shown in fig. 3.29 (a), where the input signal has a waveform corresponding to a picture on the display tube in which there are four parallel bands, alternately 'white' and 'grey', in directions at right angles to the direction of line scan. This signal can be analysed (b) into a constant d.c. component and a square-wave a.c. component. These components are passed to an amplifier through an a.c. coupling (c). The time-constant of the coupling is assumed to be sufficiently large to cause negligible distortion to the a.c. component and the drop in the d.c. component, during the time occupied by the first line of the raster, is not large enough to matter. However, the falling away of the d.c. component continues in succeeding lines so that, by the time the Nth line is reached, this component has almost disappeared. Thus the output signal from the amplifier has the form shown at (d): the first line is a nearly faithful reproduction of the input but, at the Nth line, the average level of the signal has fallen very seriously. On the display tube, the bands which should have been white will be grey, and those which should have been grey will be black. When d.c. restoration is applied (e), the d.c. component (and hence the black level) is adjusted to its correct value at the end of each line and the total output signal then becomes an almost exact replica of the input. There are several ways in which d.c. restoration may be effected and appropriate circuits are described in textbooks on television (e.g. Amos and Birkinshaw, 1962).

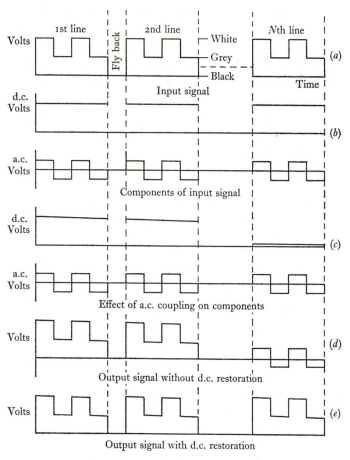

Fig. 3.29. Waveforms illustrating d.c. restoration.

3.4.2. Interrupted electron probe

The above discussion relates to conditions which obtain when, as is usually the case, the current in the incident electron probe is kept constant. Occasionally, however, (e.g. Oatley, 1969), it is advantageous to arrange for the probe current to be interrupted so that it has a square waveform, with equal 'on' and 'off' periods. When such a probe is scanned over the surface of a specimen, the waveform of the signal derived from electrons leaving the specimen will itself be basically of square waveform, but the amplitudes of the current pulses will vary from point to point. This signal can be amplified, using a.c. coupling, and d.c. restoration must then be applied so that all of the output pulses rise from a common base line, as did the input pulses. Finally, the output pulse train can be applied to a detector, which abstracts the envelope delineated by the tops of the pulses, to give a signal corresponding to variations in the properties of the specimen.

It is clear that the incident beam must be interrupted at a frequency which is several times as great as the maximum frequency of pulse-height modulation, and a factor of between five and ten is often convenient. The maximum modulation frequency is given by (3.51) in exactly the same way as for d.c. amplification, so that the above condition is equivalent to saying that the frequency of interruption must be sufficiently high to ensure that several pulses of current fall on each picture element of the specimen. For many purposes an interruption frequency of about 300 kHz is satisfactory.

The interruption of the incident electron beam could be achieved by applying a square-wave voltage between grid and cathode of the electron gun. However, since these two electrodes are commonly at a high negative potential with respect to earth, it is generally more convenient to allow the gun to supply a constant beam current and then to apply a square-wave deflecting field to sweep the beam away from the axis of the microscope and thus prevent it from passing through aligned apertures. Either an electrostatic or magnetic deflecting field may be used, but care should be taken that the beam is not swept on to an insulating surface, where the build-up of charge would create a disturbing electric field.

One obvious disadvantage of modulating the incident beam in the way described above is that the mean value of the signal current leaving the specimen is only half what it would have been if the incident beam had not been chopped. Other things being equal, therefore, the recording time must be doubled. However, for some applications this loss of current is more than compensated by other advantages.

3.4.3. Signal/noise ratio

It was shown in chapter 2 that the performance of a scanning electron microscope is limited by random fluctuations in the number of electrons falling on a picture element of the specimen surface. If n is this number, we took $\sqrt{n}$ to be a measure of the basic signal/ noise ratio (2.4) and accepted the experimental criterion that an area of brightness B cannot be distinguished from an adjacent area of brightness $B \pm \Delta B$, unless (2.5)

$$\sqrt{n} \geqslant 5B/\Delta B. \tag{3.52}$$

This equation represents a fundamental limitation which would be valid even if each incident electron produced at least one secondary and if all secondaries were collected. In practice neither of these conditions is fulfilled and, to make some allowance for this fact, we may write

$$\sqrt{n_1} \geqslant 5B/\Delta B, \tag{3.53}$$

where n_1 is the number of secondary and back-scattered electrons which leave each picture element in a single frame-scan and are collected. This equation is still not quite correct, since it makes no allowance for additional noise resulting from the randomness of the secondary-emission process itself. It will, however, suffice for our present purpose.

We shall shortly be comparing different types of amplifier and, for this purpose, it is convenient to look at the above equations from a rather different point of view. The electron current passing from the specimen to the collector is similar to that flowing in a saturated diode and we may make use of the well-known expression for the mean-square fluctuation in such a current. If I is the mean value of the current and ΔI the instantaneous departure from this mean, we have

$$\overline{\Delta I^2} = 2eI(f_2 - f_1), \tag{3.54}$$

where f_2 is the upper and f_1 the lower frequency of the band over which amplification is to be provided, while e is the electronic charge. We therefore write

$$\text{signal/noise} = I/\sqrt{(\overline{\Delta I^2})} = \sqrt{[I/2e(f_2 - f_1)]}. \tag{3.55}$$

We have already seen that f_1 is effectively zero, while f_2 is given by (3.51). Substituting these values and remembering that It/N^2e is equal to n_1, we have

$$\text{signal/noise} = \sqrt{n_1} = I/\sqrt{(\overline{\Delta I^2})} \geqslant 5B/\Delta B. \tag{3.56}$$

For some types of amplifier $\sqrt{n_1}$ is the convenient expression for signal/noise ratio while, for others $I/(\overline{\Delta I^2})$ is more useful. If we take $B/\Delta B$ to be equal to ten we have, as minimum values,

$$n_1 = 2500, \qquad (3.57)$$

$$\sqrt{n_1} = I/\sqrt{(\overline{\Delta I^2})} = 50. \qquad (3.58)$$

3.4.4. Types of amplifier

Two different types of amplifier may be used to magnify the signal obtained from the secondary current leaving the specimen in a scanning electron microscope. In the first of these the current is allowed to flow through a resistor of high value and the voltage so developed is amplified by a conventional amplifier using transistors or thermionic valves. In this case, the noise generated by the resistor and the noise produced by the amplifier itself will necessarily be added to that already present in the input signal. Thus the signal/noise ratio will be made worse by the amplification process, though we shall see later that the degradation need not be so severe that the method is useless for all purposes.

In the second type of amplifier, some physical process is used whereby each individual electron in the secondary current from the specimen produces, on the average, g new electrons, where g is greater than unity. For example, one might make use of such devices as secondary-emission multipliers, scintillators used in conjunction with photomultipliers, or multipliers depending on bombardment-induced conductivity. We deal first with amplifiers of these types.

3.4.5. Noise statistics

It might be thought that amplifiers based on such processes of direct electron multiplication would not, of themselves, introduce additional noise, but this is not the case. If it were possible to devise a process in which each initial electron produced exactly g new electrons, an amplification of g times would result and no new noise would be introduced. In practice, however, g itself is always subject to statistical variation and this brings about a degradation of the signal/noise ratio.

Following van der Ziel (1952), we consider the special case in which the noise in the electron stream entering the multiplier corresponds to random fluctuations in the mean current I, so that

$$\overline{\Delta I^2} = 2eI\Delta f, \tag{3.59}$$

where $\overline{\Delta I^2}$ is the mean-square fluctuation in a frequency band Δf.

Let β_g be the probability that an electron entering the multiplier produces g electrons in the output. Then

$$\sum_{g=0}^{\infty} \beta_g = 1; \quad \sum_{g=1}^{\infty} g\beta_g = \bar{g}; \quad \sum_{g=0}^{\infty} g^2\beta_g = \overline{g^2}. \tag{3.60}$$

Of the total current I entering the multiplier a fraction $\beta_g I$ will produce g secondaries per primary and the mean-square fluctuation in the output from this fraction will be g^2 times the corresponding input fluctuation, i.e. $2e\beta_g Ig^2\Delta f$. Taking the sum of all similar fractions, we have

$$\text{mean output current} = \sum_{g=0}^{\infty} g\beta_g I = \bar{g}I, \tag{3.61}$$

$$\text{mean-square output fluctuation} = \sum_{g=0}^{\infty} 2eI\beta_g g^2\Delta f = 2\overline{g^2}eI\Delta f, \tag{3.62}$$

$$\text{signal/noise at input} = I/\sqrt{(\overline{\Delta I^2})} = \sqrt{(I/2e\Delta f)}, \tag{3.63}$$

$$\text{signal/noise at output} = \sqrt{(\bar{g}^2 I/2e\overline{g^2}\Delta f)}, \tag{3.64}$$

and

$$\frac{\text{signal/noise at output}}{\text{signal/noise at input}} = \sqrt{(\bar{g}^2/\overline{g^2})}. \tag{3.65}$$

Thus, as a result of the variation of g, the signal/noise ratio has been made worse by the factor $\sqrt{(\overline{g^2}/\bar{g}^2)}$. A more sophisticated treatment of this problem, in which no assumption is made about the nature of the noise in the input signal, which is not necessarily given by (3.59), has been published by Shockley and Pierce (1938).

In theoretical studies, it has often been assumed that the values of β_g form a Poisson distribution so that

$$\beta_g = m^g e^{-m}/g!, \tag{3.66}$$

where m is a constant which must be chosen to give the required value for $\bar{g}$. It is easy to show that, with this distribution,

$$\bar{g} = m, \quad \overline{g^2} = m^2 + m \quad \text{and} \quad \overline{g^2}/\bar{g}^2 = 1 + (1/m), \tag{3.67}$$

so that, if $\bar{g}$ could be made sufficiently large, the multiplication process would not reduce the signal/noise ratio appreciably. It must be emphasized that this conclusion applies only to a single-stage process and would not be valid for the overall value of $\bar{g}$ produced by a cascade device such as a multi-stage electron multiplier. For a device of this kind the input noise to the second stage would not normally be given by an expression of the form of (3.59). The appropriate analysis has been given by Shockley and Pierce; it shows that if, as is usually the case, the value of $\bar{g}$ for each stage is several times as great as unity, the overall reduction of signal/noise ratio will be only slightly greater than that which would be caused by the first stage acting alone.

Even for a single-stage process, the rather meagre experimental data that are available suggest that (3.65) must be treated with considerable reserve. β_g is certainly not given accurately by (3.66) and, for the larger values of $\bar{g}$, the spread of β_g is greater than this expression would suggest. As a rough guide, we may expect that an amplifier employing direct electron multiplication by any of the usual methods will reduce the signal/noise ratio by thirty to fifty per cent.

3.4.6. The scintillator/photomultiplier amplifier

An amplifier of this type was developed by Everhart and Thornley (1960) for use in scanning electron microscopes and the details which follow are taken from their paper. The general arrangement of the apparatus is shown in fig. 3.30.

Electrons from the specimen enter a cylindrical metal box through a window of metal gauze. An accelerating potential difference of about 300 V serves to attract slow secondary electrons towards the window. Inside the box is a plastic scintillator coated with an evaporated film of aluminium about 0.07 μm thick and electrons entering through the gauze are accelerated towards this scintillator by an applied potential difference of the order of 10 kV. A metal focusing ring mounted between gauze and scintillator, serves to shape the electrostatic field in such a way that most of the electrons strike the scintillator button near its apex. The light generated in the scintillator is guided by a Perspex light pipe to a photomultiplier, which is normally mounted outside the specimen

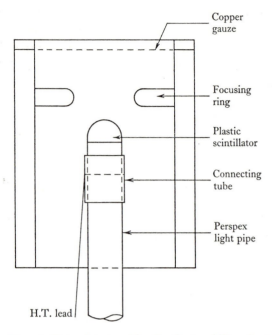

Copper
gauze

Focusing
ring

Plastic
scintillator

Connecting
tube

Perspex
light pipe

H.T. lead

Fig. 3.30. The scintillator detector. After Everhart and Thornley (1960).

chamber. The photomultiplier re-converts the light to an electron current and provides a high degree of amplification which can be controlled by variation of the potential applied to the dynodes.

In the choice of scintillator material the rate of decay of luminescence is important and, from this point of view, most of the commonly used inorganic phosphors are unsuitable. This is not always apparent from tabulated decay times, which usually refer to the afterglow experienced when the phosphor is used in a cathode-ray tube. However, for many of these materials, the decay curve is hyperbolic rather than exponential so that, at the very low light levels with which we are concerned, the decay is much slower than it is at higher brightness. For this reason one of the proprietary brands of organic plastic scintillator is commonly used.

Everhart and Thornley found that, to ensure good efficiency, considerable care was needed in machining these plastic scintillators. Their recommended procedure is to machine the material under water-cooled conditions and to polish it with a paste of French chalk

in water. Any slight heating of the surface during polishing or metal-
lizing has a deleterious effect. Their measurements gave a quantum
efficiency of about 0.02 for incident electrons with energies of 7 keV
and the efficiency rose rapidly with incident energy.

The light pipes used by these authors were made from Perspex
rods about 1 cm in diameter with polished ends. In no case did the
efficiency of transmission of light exceed sixty-five per cent and any
lack of care with surface finish caused it to fall below forty per cent.
Most of the loss was ascribed to reflection of light in entering and
leaving the pipe. Bends did not greatly impair the transmission.

From the above measurements the authors concluded that, if
electrons passing through the gauze were accelerated by a voltage
of about 8 kV before striking the scintillator, each initial electron
should, on average, cause the emission of one photoelectron from
the cathode of the photomultiplier. From our previous discussion
we should expect that noise introduced by the secondary emission
process at the first dynode of the photomultiplier would not be
negligible. In fact, the authors found that the signal/noise ratio
improved steadily as the accelerating voltage was raised from 2 to
14 kV and thereafter remained practically constant.

Plastic scintillator material that is available at the present time is
not completely satisfactory for the purpose in hand. In general,
such material is intended for the detection of nuclear particles with
energies very much higher than those encountered in a scanning
microscope. As we have seen, efficiencies for 10 keV electrons are
rather low. On the other hand, the currents to be detected in a
scanning microscope are very much higher than those normally
encountered in a nuclear detector. Perhaps for this reason, the
scintillators tend to show a falling off of efficiency with time when
used in the microscope. Materials with longer useful lives are being
actively sought.

3.4.7. Secondary-emission multipliers

Secondary-emission multipliers are widely used in a number of
sealed-off electronic devices, but the emitters commonly employed
are not stable when exposed to the atmosphere and are therefore
unsuitable for use in demountable apparatus. In some of the early
work on scanning microscopes, reported by McMullan (1953) and

by Smith and Oatley (1955), secondary-emission multipliers with beryllium–copper dynodes were used successfully. Such multipliers are unaffected by atmospheric exposure, but otherwise they compare unfavourably with detectors of the scintillator/photomultiplier type. In comparison with the latter, they introduce more noise in the multiplication process and, being rather bulky, they are much less conveniently accommodated in the specimen chamber.

Quite recently a new type of secondary-emission multiplier, the channel multiplier (Wiley and Hendee, 1962), has been developed and this device is finding application in scanning electron microscopy. The principle of the channel multiplier is shown in fig. 3.31 (a) (Adams and Manley, 1967). It consists of a glass tube whose inside wall is coated with a semi-insulating layer, so that the resistance between the ends is between 10^9 and $10^{11}\,\Omega$. The length of the tube is between 50 and 100 diameters and a potential difference of about 3 kV is applied between the ends. The device must operate in a vacuum but, with a suitable wall coating, is not harmed by exposure to air.

An electron entering the negative end of the multiplier with sufficient energy strikes the wall of the tube and liberates secondaries. These, in turn, are accelerated towards the positive end and, in due course, strike the wall again to liberate still more secondaries. So the process goes on and an overall gain of about 10^8 can be achieved before space charge causes saturation to set in. A typical curve of gain as a function of the voltage applied to the tube is shown in fig. 3.31 (b).

The secondary-emission coefficient depends on the coating material, the impact energy of the incident electrons and the direction in which they strike the surface, but an average value between 2 and 3 is to be expected for electrons with energies of 100 eV. For electrons entering the multiplier with energies of 450 eV, it has been estimated that sixty per cent give rise to chain multiplication and thus contribute to the output. On the basis of these figures, our earlier discussion suggests that the amplification provided by channel multipliers would introduce rather more additional noise than the scintillator/photomultiplier combination, but that the additional noise would not be excessive.

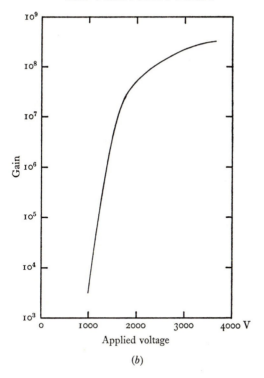

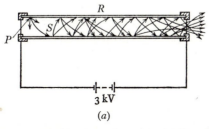

Fig. 3.31. (a) Electron trajectories in a channel multiplier. (b) Gain against applied voltage for a channel multiplier. Redrawn by permission from Adams and Manley (1967).

Although straight tubes can be used for channel multipliers, they can lead to a form of feedback. Positive ions resulting from ionization of residual gas are accelerated towards the entrance, where, if they acquire sufficient energy, they can generate unwanted electrons. Considerable energy, and thus a long free path, is needed for

8-2

this effect to be troublesome. It can therefore be cured by the use of curved tubes, which cause the ions to strike the wall before they have gained enough energy to be effective. Relatively slight curvature is sufficient, but the tubes are often coiled into spirals to conserve space. These devices may be made with input apertures ranging from several millimetres down to a fraction of a millimetre.

The use of a channel multiplier in a scanning electron microscope has been reported by Hughes, Sulway, Wayte and Thornton (1967). These authors maintained the input end of the multiplier at a positive potential of between 12 and 250 volts with respect to the specimen and the output signal from the multiplier was fed to a conventional d.c. amplifier, and from thence to the display unit. If the output signal from the multiplier had been taken from its positive end, it would have been necessary to keep the d.c. amplifier at a potential of several kilovolts with respect to earth or to provide a balancing voltage. To avoid this difficulty, the output signal from the multiplier was taken from its negative end, while a $10^8 \, \Omega$ resistor was connected between the positive end of the multiplier and the positive terminal of its high-tension supply unit (fig. 3.32). With this arrangement, the high resistance from end to end of the multiplier was effectively in series with the output of the multiplier, and the frequency response of the amplifier was thus limited to about 10 kHz. With a total scan time of 200 seconds, good micrographs were obtained with an incident probe current of 10^{-12} A. It was found desirable to insert an earthed metal screen between the specimen and the bulk of the multiplier, to prevent the occurrence of a large electrostatic field at the specimen. The low-potential end of the multiplier was allowed to project through a hole in this screen.

An alternative method of using a channel multiplier in a scanning microscope would be to place the load resistor at the output end of the multiplier and to emply a.c. coupling with d.c. restoration. Isolation of the high potential of the output end of the multiplier from the remainder of the circuit would then be achieved by the coupling capacitor. With this arrangement, the effective output resistance of the multiplier would be greatly reduced and the frequency response correspondingly improved. The upper limit of frequency of the channel multiplier itself, when not reduced by coupling circuits, is of the order of 10 Mc/s.

Since the length of a channel multiplier must be large in comparison with its diameter, the input aperture of a single multiplier, which can be accommodated within the specimen chamber, is necessarily rather small. For some purposes this may be an advantage while, for others, a detector of larger area is needed. The latter requirement may be met by channel plates which are now becoming available. These consist of bundles, containing large numbers of channel multipliers, fused together, the surfaces of the plates being ground and polished.

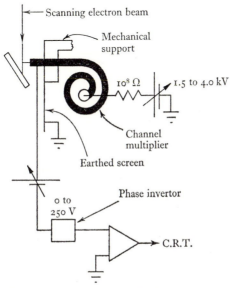

Fig. 3.32. The use of a channel multiplier in a scanning microscope. Redrawn by permission from Hughes *et al.* (1967). *J. App. Phys.* **38**, 4922.

3.4.8. Semiconductor detectors

When electrons with sufficiently high energy enter a semiconductor, each electron produces several hole/electron pairs. If electrodes are attached to opposite faces of a slice of the semiconductor and are connected respectively to the positive and negative terminals of a battery, the holes and electrons produced by the incident electron stream flow in opposite directions and a current is generated in the external circuit. Devices of this type have been extensively studied as detectors of nuclear particles and, more recently, they have been

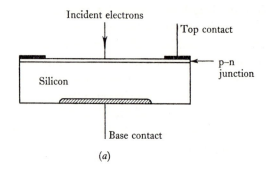

(a)

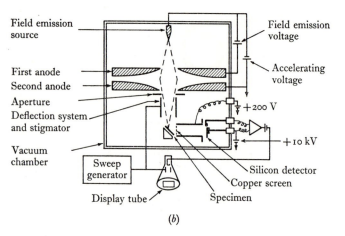

(b)

Fig. 3.33. (a) Diagram of a semiconductor detector. (b) The use of a semiconductor detector in a scanning microscope. Redrawn by permission from Crewe, Isaacson and Johnson (1970).

applied to the detection of the secondary-electron current in a scanning electron microscope. Detailed information about the underlying theory and the methods of construction of efficient detectors has been given by Dearnley and Northrop (1963) and we shall here be concerned only with general principles.

The form of detector most suitable for our present purpose is illustrated in fig. 3.33 (a). A thin slice of high-resistivity silicon has a p–n junction formed just below its top surface. This may be achieved by allowing a suitable impurity to diffuse downwards through the top surface. Alternatively, if the original material is n-type, an inversion layer may be produced near the upper surface by the 'surface barrier' technique, in which the material is first cleaned and etched and is then allowed to oxidize by standing in air at room temperature. A metallic ohmic contact is made to the lower face and the top face is coated with a thin evaporated layer of gold which forms the top contact. The incident particles enter the semiconductor through this layer.

The p–n junction can be made to serve in one of two ways. If an external voltage is applied to the device, as has been indicated above, a current flows even when no incident electrons are falling on the device. This 'dark' current generates noise and so must be kept as small as possible. If the polarity of the external voltage is such that the rectifying junction is reverse-biased, the dark current will be much smaller than it would have been in the absence of the junction and, when incident electrons fall on the detector, the signal/noise ratio will be correspondingly better.

Alternatively, it is possible to use the detector without any external bias, since the p–n junction provides a built-in electric field which serves to separate any holes and electrons which are generated in the depletion region. The dark current is then reduced to zero and this mode of operation affords the best signal/noise ratio that can be obtained with a given device.

The presence of the p–n junction is not without its disadvantage, since it adds considerably to the effective capacitance in parallel with any load resistor in the external circuit and thus reduces the speed of response of the detector. The capacitance C per unit area of a p–n junction, which is reverse-biased by external voltage V, obeys the equation

$$C \propto (V + V_\mathrm{d})^{-\frac{1}{2}}, \tag{3.68}$$

where V_d is the built-in voltage step at the junction itself. V_d depends on impurity concentrations, but is of the order of 1 V for silicon. It thus appears that the condition $V = 0$, which gives the best signal/noise ratio, is also the condition which leads to the maximum value of C. Thus a compromise must be reached between signal/noise ratio and speed of response.

The use of a semiconductor detector in a scanning electron microscope has been described by Crewe, Isaacson and Johnson

(1970). The arrangement used by these authors is shown in fig. 3.33(b). Secondary electrons from the specimen were attracted towards a copper gauze closing the end of a beryllium–copper tube, which was maintained 200 V positive with respect to the specimen. A further accelerating voltage of up to 10 kV was applied between the other end of the tube and the silicon detector. The current generated by the detector was led through the wall of the specimen chamber to a load resistor and the resultant voltage drop was amplified by a battery-driven transistor amplifier. The output of this amplifier was a.c. coupled through a capacitor to the display unit. For the purpose in hand, d.c. levels of the secondary-electron signal were not wanted, so it was considered unnecessary to provide d.c. coupling throughout or d.c. restoration.

The silicon detector was about 0.5 cm square and was operated without external bias. Its capacitance was then about 350 pF and a load resistor of 100 kΩ was found suitable. The gold film on the front face of the detector was made as thin as possible (less than 0.02 μm) and it was estimated that each incident electron produced about 2000 electron/hole pairs. The output signal/noise ratio obtained with the combination of semiconductor detector, coupling circuit and transistor amplifier was estimated to be about thirty times worse than that to be expected from a scintillator/photomultiplier detector. It was felt, however, that the system could have been improved by using an amplifying system with lower noise.

3.4.9. Conventional amplifiers

We turn now to the alternative procedure in which the signal current from the specimen is allowed to flow through a resistor and the voltage so developed is amplified by conventional means. For the production of the initial voltage signal there are the two possibilities shown in fig. 3.34. In fig. 3.34(a) the current which flows through the resistor R is the difference between the incident beam current and the current which leaves the specimen by way of secondary or back-scattered electrons. Since the former component is constant, the signal voltage is derived from the total electron current which leaves the specimen. This current will be influenced by the potential of the specimen with respect to its surroundings; for example, by making the specimen sufficiently positive, the slow-secondary electrons can be prevented from leaving it. Thus it is possible to include or to exclude these secondaries from the signal that is passed to the amplifier. An advantage of this arrangement is that electrons leaving the specimen in all directions contribute to the signal, but there is the converse disadvantage that effects which depend on deriving the signal from electrons leaving in particular directions cannot readily be studied. There is one further disadvantage which may sometimes be important. If conditions are

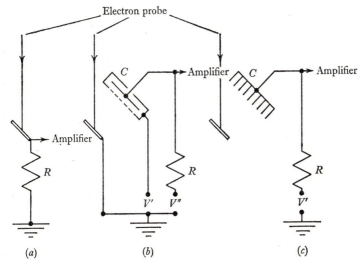

Fig. 3.34. Arrangements for signal detection using conventional amplifiers.

such that the current leaving the specimen is small compared with the incident beam current, the signal/noise ratio will be poor; the major part of the incident current will contribute to the noise but not to the signal.

In the alternative arrangements of fig. 3.34(b) and (c), the signal is derived from those electrons which have left the specimen and which enter a suitable collector C. Many of the electrons leave the specimen with high energy and, when these impinge on C, they will themselves produce secondaries which must be retained if there is to be no loss of signal. To achieve this end, C can take many forms, of which two are shown. In fig. 3.34(b) the resistor is connected to a metal plate which is maintained at a positive potential V'' of about

100 V with respect to a surrounding metal box which the electrons enter through a metal gauze window. The box itself may be maintained at any suitable potential V' with respect to the specimen to attract or repel slow secondaries as desired. In fig. 3.34(c) the surface of C facing the specimen has a structure resembling a honeycomb, so that incident electrons have a high probability of entering one of the cells. Most of the secondary electrons are then trapped because they strike the cell walls before they can escape.

The type of amplifier to be used to magnify the signal voltage developed across the resistance R is governed by the following considerations. We have already seen that the signal current I reaching the collector C has superimposed on it a noise current. Part of this noise is the result of the random arrival at the specimen of electrons in the incident probe and this component has mean-square amplitude (§3.4.3)

$$\overline{\Delta I^2} = 2eI\Delta f. \tag{3.69}$$

Additional noise may also result from the random variation of the secondary-emission coefficient at the specimen surface, but the extent of the addition is usually unknown and, for the approximate calculation that follows, we shall ignore this effect completely.

The noise current given by (3.69) produces across the resistance R a mean-square noise voltage

$$\overline{\Delta V_1^2} = 2eIR^2\Delta f, \tag{3.70}$$

and to this must be added the Johnson noise generated by R itself. This latter is given by

$$\overline{\Delta V_2^2} = 4kTR\Delta f, \tag{3.71}$$

where T is the absolute temperature and k is Boltzmann's constant.

Our earlier discussion of the various types of amplifier using some form of direct electron multiplication has shown that the best of these is likely to cause some degradation of the signal/noise ratio, which may amount to 30–50 per cent. If the scheme of fig. 3.34 followed by a conventional amplifier is to give a performance at all comparable with the multiplier types, it is clear that, as a first requirement, the noise represented by (3.71) must not exceed that given by (3.70). A second requirement is that the amplifier which follows should not degrade the signal/noise ratio appreciably.

The first condition gives

$$2eIR^2\Delta f > 4kTR\Delta f, \tag{3.72}$$

or
$$R > 2kT/eI. \tag{3.73}$$

At room temperature, this reduces to

$$R > 5.2 \times 10^{-2}/I. \tag{3.74}$$

If, therefore, we wish to work with mean signal currents of the order of 10^{-11} A, R should not be much less than $10^{10}\,\Omega$. The use of a resistor of such high value imposes considerable limitations on the design of the amplifier which is to follow it and little work appears to have been done to determine the best type of amplifier. Moreover, the possibilities are continually being extended by the introduction of new transistors with improved characteristics. For these reasons we shall not consider details of design, but shall content oureslves with general principles which appear to be relevant.

There will be stray capacitance between the electron collector and earth and this, added to the input capacitance of the amplifier itself, with which it is in parallel, gives a total which is unlikely to be much less than 5 pF and may be much greater. The shunting effect of this capacitance across a resistor of $10^{10}\,\Omega$ severely attenuates all but the lowest frequencies, so the amplifier must be designed to provide appropriate correction. This is usually done by the inclusion of negative feedback, but that in itself is not entirely straightforward. The $10^{10}\,\Omega$ resistor will have self capacitance across its terminals and this must be taken into account in designing the feedback network. A method of overcoming these problems was suggested by Pelchowitch and Zaalberg van Zelst (1952), Van Nie (1968) and others. These authors were concerned with the amplification of currents of the order of 10^{-15} to 10^{-13} A with band-widths up to 300 Hz. There appears to be no reason why their methods should not be extended to larger currents and larger band-widths, as required for scanning microscopy, but further work is clearly needed. The general principles of the circuit which they used are discussed below.

In fig. 3.35, the input signal is derived from the constant-current generator, which sends current through the resistor R which has a resistance of the order of $10^{10}\,\Omega$. C is the self-capacitance across

R and C_1 is the sum of the output capacitance of the current generator and the input capacitance of the amplifier. The amplifier has voltage gain A, which is assumed to be constant over the frequency range considered, and its input resistance is taken to be infinite; with field-effect or MOS transistors, these assumptions are close approximations to the truth. The impedance of the feed-back network consisting of R_2 in series with C_2 is large compared with the output impedance of the amplifier, but small in comparison with the impedance of the parallel combination of R and C. Again, there is no difficulty in satisfying these conditions. If R_2 and/or C_2 is chosen to satisfy the condition

$$R_2 C_2 = RC, \tag{3.75}$$

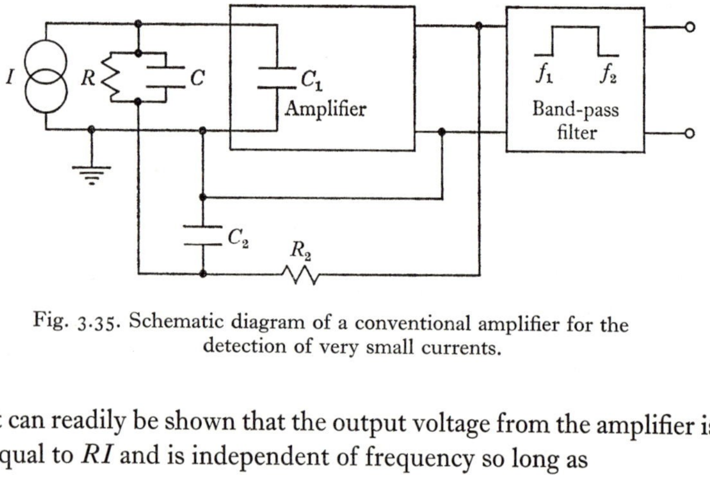

Fig. 3.35. Schematic diagram of a conventional amplifier for the detection of very small currents.

it can readily be shown that the output voltage from the amplifier is equal to RI and is independent of frequency so long as

$$A \gg R(C + C_1)\omega. \tag{3.76}$$

The output from the amplifier is taken through a band-pass filter to exclude frequencies which lie outside the wanted band: in this way noise is reduced to a minimum.

We have already discussed the components of noise arising respectively from the randomness of the electron stream in the input current I and from Johnson noise in the resistor R. In addition there will be noise contributed by the amplifier and this, in turn, contains two components. The first of these is independent of frequency and can be represented by an equivalent noise resistance R_n in series with the input to the amplifier. The second component, sometimes termed 'flicker noise' or '$1/f$' noise, produces a mean-square voltage at the input to the amplifier which is inversely proportional to frequency. The total amplifier noise, expressed as a voltage ΔV_3 at the input terminal, is therefore given by

$$\overline{\Delta V_3^2} = 4kTR_n(1 + f_n/f)\Delta f, \tag{3.77}$$

where Δf is a small frequency interval and f_n is the frequency at which the flicker noise is equal to the other component represented by R_n. Both R_n and f_n depend principally on the type of transistor or valve used in the first stage of the amplifier.

It is of interest to compare the amplifier noise with the Johnson noise introduced by the resistor R and, for this purpose, it is convenient to refer both components of noise to the input terminals of the amplifier. To do this we note that the series resistive component of the impedance of the input circuit, as seen from the terminals of the amplifier, is not R but $R/[1 + R^2(C+C_1)^2\omega^2]$ if the impedance of the feedback loop is neglected. Hence the mean-square noise voltage at these terminals, resulting from this component is given by

$$\overline{\Delta V_4^2} = 4kTR\Delta f/[1 + R^2(C+C_1)^2\omega^2], \tag{3.78}$$

and it is this quantity which must be compared with $\overline{\Delta V_3^2}$ (3.77). In passing, we note that the shunting effect of $(C+C_1)$ on the noise reaching the amplifier from R does not invalidate our previous conclusion (3.74) as to the value of R needed to ensure that the Johnson noise does not greatly exceed the shot noise inherent in the input current I. Both of these components are affected in the same way by the capacitative shunt.

We can now compare the relative contributions of amplifier noise and Johnson noise, for any small frequency interval by the ratio

$$\frac{\overline{\Delta V_3^2}}{\overline{\Delta V_4^2}} = \frac{R_n}{R}\{(1+f_n/f)[1 + R^2(C+C_1)^2 4\pi^2 f^2]\}$$

$$= \frac{R_n}{R}\left[1 + \frac{f_n}{f} + 4\pi^2 R^2(C+C_1)^2(ff_n+f^2)\right]. \tag{3.79}$$

To get the average ratio over the whole of the required frequency band from f_1 to f_2, we integrate (3.79) and divide by (f_2-f_1) to give

$$\left[\overline{\frac{\Delta V_3^2}{\Delta V_4^2}}\right]_{\text{Average}} = \frac{R_n}{R(f_2-f_1)}\left\{(f_2-f_1)+f_n\ln\frac{f_2}{f_1}\right.$$
$$\left.+4\pi^2R^2(C+C_1)^2[\tfrac{1}{2}f_n(f_2^2-f_1^2)+\tfrac{1}{3}(f_2^3-f_1^3)]\right\}. \quad (3.80)$$

Substitution into this expression of appropriate values for f_1, $f_2(C+C_1)$, R and R_n suggests that, for many of the conditions encountered in a scanning electron microscope, the amplifier might increase by a small factor the noise already existing in the input current I. Experimental work to test this conclusion is at present in hand.

3.4.10. General review of detectors

At the present time, the plastic scintillator plus photomultiplier is the type of electron detector most commonly used in scanning microscopes. It provides a better signal/noise ratio than any other type that has so far been proposed and seems likely to retain a good deal of its popularity in the future. On the debit side, it is expensive, rather bulky and not too convenient to manipulate inside the specimen chamber. If an ultra-high vacuum were needed in the specimen chamber, the presence of a plastic scintillator and a Perspex light pipe would be objectionable. This difficulty might be overcome by the use of an inorganic scintillator and a glass light pipe, but data on any consequent changes in the efficiency of the detector do not appear to have been published.

When some degradation of signal/noise ratio can be tolerated, other types of detector may well be considered but, of these, the channel multiplier seems to have little advantage over the scintillator–photomultiplier combination, unless electrons are to be collected from a small and rather inaccessible area of the specimen. Channel multiplier plates, of large area, are becoming available, but seem likely to be expensive.

Semiconductor detectors are cheap and appear to have many advantages when the introduction of some additional noise is permissible. This may be the case when a field-emission cathode is used (Crewe, Isaacson and Johnson, 1970) or when the operation of the microscope is to be confined to relatively low magnification and a fairly large incident probe current can be used.

The combination of a metallic electron collector with a conventional transistor amplifier has, so far, received little attention, but may yet prove valuable. It differs in one respect from all other detectors in that, when it is used to abstract a signal from the total current flowing between specimen and earth (fig. 3.34(a)), it effectively records all electrons leaving the specimen, since the incident current is constant. This may, or may not, be what is required, but it is something which no other detector is able to do.

3.4.11. Other methods of signal generation

The methods described above are quite general ones, which can be used with any specimen to generate a signal which effectively depends on the number of electrons collected from each point of that specimen. We turn now to quite different methods of deriving a signal, most of which are not of universal application.

The most important of these is the technique which is known as X-ray microanalysis. When the incident electron probe strikes the specimen, some of the electron energy is used in generating characteristic X-rays of the atoms present in the material. These X-rays can be allowed to pass through a crystal spectrometer to a proportional counter, or to a solid-state detector with pulse-height analyser, so that an analysis can be made of the atoms from which the X-rays originated. In this way, a picture of the distribution of elements in the specimen can be built up as the incident probe is scanned over its surface. Additional information can then be obtained by comparing this picture with the corresponding one obtained by the collection of electrons leaving the specimen, as in the conventional scanning microscope.

It is clear that the X-ray microanalyser and the scanning electron microscope have much in common and that a single intrument can serve both functions. There are, however, important differences of technique. The generation of X-rays is an inefficient process and, of the X-ray photons generated, only a fraction reach the detector. It follows that, for a given incident probe current, the number of detected photons per picture point in the microanalyser is much smaller than the corresponding number of electrons collected per picture point in the scanning microscope. Noise is therefore a much more serious problem in the microanalyser and, to obtain satisfactory results, the incident beam current must be larger and poorer resolution must be tolerated. Because of this basic difference and because the accommodation of the X-ray spectrometer and detector raises special problems, the design of an instrument built primarily for microanalysis is like to differ in many respects from a scanning microscope. We shall not consider these matters further; the X-ray microanalyser is an important instrument in its own right and merits separate treatment. A review of recent developments in X-ray microanalysis has been given by Duncumb (1969).

With certain specimens, the incident electron probe gives rise to the generation of ultraviolet, visible or infrared radiation and, as with X-rays, this radiation can be detected and used to provide a signal from which a 'picture' of some property of the specimen can be built up. Clearly, the technique is applicable to only a very restricted range of specimens, notably phosphors and certain semiconductors, and for this reason we shall not discuss it further. A review of the subject has been given by Thornton (1968).

When the specimen is an insulator or a semiconductor, the incident electron probe will usually generate electron/hole pairs in the material. Since the energy required to produce such a pair is only a few electron volts, each incident electron may produce a considerable number of pairs. If an electric field exists within the material, holes and electrons will be separated and, if an external circuit is connected to the specimen via electrodes attached at suitable places, current will flow in this circuit. The current can clearly provide a signal which may be passed to the display unit, but the interpretation of the resulting 'picture' will often be a matter of extreme difficulty. The electric field in the specimen may be produced by bias applied by the external circuit, but it may also be caused by a p–n junction or other barrier within the specimen itself. The free motion of the holes and electrons may be modified by the existence of dislocations or grain boundaries, by recombination, or by the build-up of space charge as a result of trapping. Unless one has a great deal of information about the specimen being examined, it is likely to prove very difficult to disentangle these various effects. Moreover, much of the action will occur at distances below the specimen surface which are of the order of the range of the incident electrons in the material. These matters, also, have been discussed by Thornton (1968).

If the techniques in this section have been discussed somewhat briefly, that is not because they are unimportant. X-ray micro-analysis has already proved to be of the highest value, and the other methods of deriving signals may yield information of great utility in particular cases.

3.5. The vacuum system

In any electron microscope the pressure must be reduced to a value such that most of the electrons complete their trajectories before colliding with gas molecules and, so far as this criterion is concerned, a pressure between 10^{-4} and 10^{-5} Torr is satisfactory. On the other hand, if a field-emission cathode is to be used, an ultra-high vacuum ($\sim 10^{-10}$ Torr) must be maintained in the electron gun; with certain specimens it may also be desirable to reduce the pressure to a similar value in the specimen chamber to prevent contamination. Between these limits a compromise must be reached whereby the pressure is kept sufficiently low for the purpose in hand without complicating the pumping system unnecessarily. The means by which these pressures are achieved are those commonly employed in vacuum practice and only the following points call for special mention.

If a scanning microscope is to give high resolution it must be shielded, as far as possible, from all sources of vibration. Thus, when a mechanical backing pump is used, it should preferably be situated at a distance from the microscope and mounted on antivibration supports. Similarly, vibration may sometimes be caused by the boiling of oil in a diffusion pump, which is usually mounted close to the microscope column to avoid reduction of pumping speed. For this reason, an ion pump is cometimes preferred to a diffusion pump, even when ultra-high vacuum conditions are not needed.

Oil vapour from a backing pump or a diffusion pump, which reaches the specimen chamber, will be adsorbed on the specimen surface and subsequently decomposed by the electron probe. If the adsorbed layer is continuously replenished, contamination can build up quite rapidly and a deposit 0.01 μm thick can be formed in a few minutes, the time depending on the electron beam current. It is therefore worth while to go to considerable lengths to prevent oil vapour from reaching the specimen chamber. A trap may be inserted between the mechanical backing pump and the diffusion pump and a thermo-electrically-cooled baffle between the latter and the specimen chamber. If necessary, the specimen itself may be surrounded by a cooled baffle. These matters have been studied by Ennos (1953, 1954), who has also investigated the contamination arising from vacuum grease, gasket materials, uncleaned metal surfaces and other substances likely to be found in a vacuum system.

For many purposes it is convenient to make provision for the specimen to be introduced into the microscope through an air-lock, so that the consequent rise in pressure is kept to a minimum. On the other hand, if the specimen is to be subjected to electrical or mechanical changes while under observation, it can often with advantage be set up in a large specimen chamber, with facilities for its subsequent processing. A pump-down time of five or ten minutes may then have to be accepted.

3.6. The display unit

In a scanning electron microscope the final magnified image is formed on a cathode-ray tube and it is customary to provide at least two separate display channels for visual observation and for photographic recording respectively. The visual tube will usually be scanned with frame periods of the order of a second and a screen with relatively long afterglow greatly reduces flicker. Such screens generally give an initial short flash of blue light, followed by the afterglow in the yellow or red part of the spectrum. The sensation of flicker is lessened if the brightness of the initial flash is reduced, and this is achieved by viewing the screen through a suitable optical filter. In a cathode-ray tube of this kind, the light (or ultraviolet) produced by the initial impact of electrons on the screen excites afterglow in the surrounding phosphor. Thus the effective spot size is larger than the diameter of the electron beam and the resolution of the tube is unlikely to be very high. A typical value would be 500 lines in a square screen of side ten centimetres and this is satisfactory for visual observation.

For photograpnic recording much higher resolution is desirable and a tube giving 800–1000 lines in a 10 cm square is commonly used. In this case, a phosphor with very short afterglow is chosen.

We have seen that, at high magnifications, the recording time may be as long as 100 seconds or more, in order to allow a sufficient number of electrons to fall on each element of the specimen. In principle, it is immaterial whether the total recording time is occupied by a single scan or by a number of repeated scans, but the former alternative is greatly to be preferred. If there is any slow drift in the position of the specimen or in the accelerating voltage applied to the incident probe, repeated scans will not be exactly superimposed and the image recorded photographically will thus be blurred. With a single long-period scan, under the same conditions, the photographic record will remain sharp and the slight distortion resulting from the drifts will often be harmless.

The accelerating voltages applied to the cathode-ray tubes must be stabilized; a stability of 0.1 per cent is usually adequate and is easy to provide.

3.7. The complete instrument

The principles which must govern the design of the component parts of any scanning microscope have been explained in some detail in the foregoing sections. Within the limitations set by these principles the designer has considerable latitude in his choice of components and in the manner in which he assembles them to form the complete instrument. In this section we draw attention, without detailed discussion, to some further matters to which he will need to give consideration.

(a) The number of lenses

For operation at high magnification, the diameter of the incident electron probe should be 0.01 μm or less. If, as is commonly the case, the electron gun makes use of a tungsten-hairpin cathode, the diameter of the cross-over will be somewhat less than 100 μm, so an overall demagnification of at least 10 000 times is needed. The focal length of the final lens cannot be much less than 1 cm if an adequate working distance is to be available, so the required demagnification necessitates the use of two earlier lenses, if the total length of the column is to be kept within reasonable bounds. For a general-purpose instrument, three lenses in all are commonly provided.

In certain circumstances, however, fewer lenses will suffice. If a pointed cathode is used, the effective size of the source is likely to be very much less than 100 μm and it may then be possible to produce the required demagnification with two lenses or even with a single lens.

(b) Adjustments to the triode electron gun

When a new cathode is inserted in an electron gun it is rarely possible, at the outset, to ensure that the electron beam directed along the axis of the column has its maximum intensity. Even if this were possible, some drift in the positions of the electrodes might well occur during operation of the microscope. It is therefore convenient to make provision for some adjustment to the position of the cathode or other electrode to be effected by controls located outside the vacuum system or for the beam to be deflected by an electric or a magnetic field. The optimum position can then be found by trial.

If the gun is to be operated over a wide range of accelerating voltage, the distance between cathode and anode should be variable. To overcome space charge, we wish to establish at the cathode surface the maximum electric field consistent with freedom from breakdown. This will clearly entail different spacings between anode and cathode at different voltages.

(c) The specimen stage

To enable the operator to examine different areas of a specimen, the latter should be movable in two mutually perpendicular directions normal to the incident electron beam. Ideally, he would also like to be able to rotate the specimen and to vary its inclination to the electron beam and its distance from the final lens. However, to provide all these motions, with controls outside the specimen chamber is a matter of considerable complexity and, in the interest of simplicity, a designer will often dispense with some of them.

The basic X–Y movements are usually provided by micrometers acting through O-ring seals and, since fine adjustments are needed, the micrometers should be of high quality. For still finer control of the area of specimen under observation, separate windings may be added to the scanning coils. Variable direct currents through these coils may then be used to change the area scanned by the electron beam.

Small electric motors, mounted inside the specimen chamber, have also been used to provide some of the movements required.

(d) Vibration

At high magnification, any vibration of the microscope can seriously impair the quality of a micrograph. This is particularly the case when a field-emission cathode is used, since there is then little or no demagnification of the source; any vibration of the cathode causes a roughly equal shift of the point in which the incident probe strikes the specimen.

In many buildings the floor has a dominant frequency of vibration of a few hertz. It is therefore good practice to mount the microscope on elastic supports, so that its natural frequency of vibration is not greater than one hertz and preferably a good deal less. Several types of anti-vibration support are available for this purpose.

CHAPTER 4

THE INTERACTIONS OF ELECTRONS
WITH A SOLID

4.1. Introduction

In the scanning microscope an incident pencil of electrons, of relatively high energy, strikes the surface of the solid specimen under observation; thus the way in which electrons interact with a solid is of great interest. Some electrons penetrate the solid to depths of the order of 1 μm or, if the specimen is sufficiently thin, pass through it with reduced energy and changed direction. Others are turned back and leave by the surface through which they entered, often without great loss of energy. Some of the primary electrons excite secondaries which leave by the surface on which the incident beam impinged. Any or all of these processes may play a part in the formation of the image which appears on the cathode-ray tube, so a general understanding of their nature and of the laws to which they conform is essential to an intelligent use of the scanning electron microscope.

A great deal of theoretical and experimental effort has been devoted to problems of the interaction of electrons with solids, but much of it is irrelevant to our purpose. At the present time, no theory of the interaction is completely satisfactory so we shall confine ourselves largely to a discussion of experimental results. Again, much of the experimental work has been concerned with electrons of such high energy that they are unlikely to be used in scanning microscopes. Our object will be to present a general picture of the processes mentioned above, without going into too much detail. We shall draw largely on results published by Cosslett and Thomas (1964a, b, c, and 1965). These authors worked with electrons in the energy range 5–30 keV and have discussed very fully the relation of their own results to those obtained by other observers. They also provide comprehensive lists of references and their papers form an excellent starting point for anyone who wishes to pursue the subject further.

4.1.1. Physical basis of electron scattering

When an electron collides with an atom of a solid, the electron may be scattered elastically or inelastically. So far as the incident electron is concerned, inelastic collisions are generally unimportant, since elastic collisions occur much more frequently. Inelastic collisions may have an appreciable effect on the scattering of the incident electron in elements of atomic number lower than about six; more important for our present purpose is the fact that they may lead to the production of secondary electrons, which we shall consider later.

Dealing first with a single scattering event, the incident electron may pass close to the atomic nucleus. The orbital electrons then have little effect and the scattering of the incident electron, usually through a large angle, can be calculated from Rutherford's original expression. Alternatively, the incident electron may pass through the outer part of the atom where the field of the nucleus is screened to a greater or less extent by the orbital electrons. In this case, any calculation of scattering must depend on the model chosen to represent the way in which the total electrostatic field varies with distance from the centre of the atom. A model introduced by Wentzel (1927) has been very generally used and from it Cosslett and Thomas (1964a) deduce an expression for the number of elastic collisions p_e which occur as an incident electron of energy E_0 (keV) transverses a film of thickness x (cm) and density ρ (g/cm³). If Z is the atomic number and A the atomic weight of the material,

$$p_e = 3 \times 10^6 Z^{\frac{4}{3}} \rho x / A E_0. \tag{4.1}$$

This expression assumes that the energy of the electron does not change appreciably as it traverses the film; when x is large enough to invalidate this assumption, a correction must be made. Moreover x is properly interpreted as path length rather than film thickness.

These authors also use the Wentzel model to calculate the ratio of the number of elastic collisions $p(\theta)$, which result in a deflection of the electron through an angle greater than θ from its original direction, to the total number of elastic collisions p_e. Their results are reproduced in table 4.1.

Too much reliance must not be placed on the accuracy of (4.1) or the figures in table 4.1, since both rely on the Wentzel model, which cannot be more than an approximation to the truth. Nevertheless they provide us with a general picture of the facts about single scattering, which cannot be very greatly in error. In particular, we note that the fraction of elastic collisions resulting in appreciable deflection rises as Z is increased and as E_0 is decreased.

TABLE 4.1. *Ratio of the number of elastic collisions $p(\theta)$ with a deflection greater than θ to the total number of elastic collisions p_e*

E_0	Element	$p(10°)/p_e$	$p(30°)/p_e$	$p(90°)/p_e$
	Al	0.10	0.013	0.0009
20 keV	Cu	0.18	0.023	0.0017
	Au	0.28	0.042	0.0030
	Al	0.20	0.026	0.0019
10 keV	Cu	0.31	0.050	0.0036
	Au	0.45	0.089	0.0064
	Al	0.31	0.050	0.0036
5 keV	Cu	0.46	0.096	0.0068
	Au	0.61	0.168	0.012

As the incident electron penetrates into the solid it suffers a number of elastic collisions of the type we have been discussing and, so long as this number remains small, we speak of *plural scattering*. However, when the number reaches about twenty-five, we can apply error theory to the problem and the electron is then said to have entered the region of *multiple scattering*. In this region, Bothe (1921, 1933) has suggested that the fraction of the incident current scattered into unit solid angle making angle θ with the original direction should be given by

$$N(\theta) = (2\pi\lambda^2)^{-1}\exp\left(-\theta^2/2\lambda^2\right), \qquad (4.2)$$

where λ is the most probable angle of total deflection. It will, of course, depend on the nature of the scattering material and on the thickness of material traversed. Since the solid angle lying between θ and $(\theta+d\theta)$ is $2\pi\theta.d\theta$, the fraction η_θ of the incident pencil of electrons which, after scattering, lies within a cone of semi-angle θ is by (4.2),

$$\eta_\theta = \int_0^\theta N(\theta)\,2\pi\theta.d\theta = 1 - \exp\left(-\theta^2/2\lambda^2\right). \qquad (4.3)$$

Experimental results substantiate Bothe's theory to the extent that a Gaussian expression of the form of (4.3) is found to hold. However, they are not in agreement with his expression for λ, possibly because he neglects the effect of large-angle single scattering.

As the incident electron beam traverses still greater thicknesses of scattering material and the number of elastic collisions gets large, the motion of individual electrons becomes completely random and the conditions are said to correspond to *diffusion*. A different form of theory is then appropriate. When these conditions apply, theory suggests that the fraction η_D of incident electrons which are transmitted through a thickness x of scattering material is given by an expression of the form

$$\eta_D = \exp(-\mu x). \tag{4.4}$$

This is in agreement with experiment, but attempts to derive the value of μ in terms of fundamental atomic constants have not been very successful.

Another feature of the diffusion regime is that the most probable angle of total deflection, which increases steadily with depth of penetration so long as conditions of multiple scattering obtain, reaches a limiting value when diffusion sets in. This fact, as well as the validity of (4.4), can be used as a criterion of the onset of diffusion when analysing experimental results. The minimum thickness needed to produce diffusion is known as the *diffusion depth* x_D.

In the following sections we shall review the experimental evience in the light of the above picture of physical processes. Almost all experiments have been carried out with films of metal, because this avoids trouble resulting from charging of the film. Apart from this trouble, there is no reason to suppose that non-metallic films would behave differently from those of metal.

4.1.2. Experimental results on transmission of electrons: Cosslett and Thomas (1964 b)

The fraction η_T of electrons, incident normally on a film, which succeed in traversing a thickness x is shown in figs. 4.1 (a) and (b) for copper and gold respectively. The product of the thickness x and density ρ of the material (the mass-thickness) is used as a parameter rather than x itself because, when plotted in this way, the two sets of curves show considerable similarity, despite the wide difference in atomic number between copper and gold. Curves for aluminium

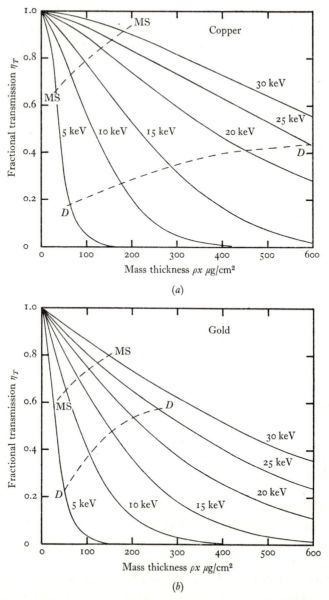

Fig. 4.1. The fractional transmission of electrons incident normally on a film, as a function of incident energy and of film thickness. After Cosslett and Thomas (1964b).

are similar to those for copper, while curves for silver are inter-
mediate between those for copper and gold respectively.

Using the criteria previously discussed, dotted lines are drawn
across the curves to indicate the onset of multiple scattering (MS)
and diffusion (D) respectively. Once the thickness is great enough
for diffusion conditions to have been established, (4.4) applies and
we may write
$$\eta_T = \eta'_T \exp(-\mu x'),\qquad(4.5)$$
where η'_T is the value of η_T at the diffusion depth x_D where conditions change from multiple scattering to diffusion and x' is the
additional thickness measured from this depth (i.e. $x' = x - x_D$).
Values of the absorption coefficient depend greatly on the material
and it is usual to tabulate the *mass absorption coefficient* μ/ρ which
does not vary greatly with Z. Values of this quantity for copper,
silver and gold, are given in table 4.2 for several values of E_0. As
usual, E_0 is the energy of the electrons when they first impinge on
the material; it is not the energy which they have when diffusion
conditions have been established.

TABLE 4.2. *Mass absorption coefficients*

E_0 (keV)	5	10	15	20	25	
Cu	4.0	1.5	0.78	0.25	0.16×10^4	
Ag	4.0	1.4	0.73	0.33	0.21×10^4	μ/ρ (cm^2/g)
Au	4.0	1.4	0.72	0.40	0.25×10^4	

4.1.3. Experimental results on the angular distribution of scattered electrons: Cosslett and Thomas (1964 *b*)

When a narrow beam of electrons of initial energy E_0, is incident
normally on a thin film of material, the transmitted electrons will
be scattered in various directions. We denote by $N(\theta)$ the fraction of
the incident electrons which are scattered into unit solid angle
making angle θ with the original direction. Then the total fraction
scattered into directions making angles between θ and $\theta + d\theta$ with
the normal is given by $N(\theta) 2\pi \sin\theta . d\theta$. Curves of $N(\theta)$ and of
$N(\theta) 2\pi \sin\theta$ for various thicknesses of copper film and for E_0 equal
to 20 keV are given in fig. 4.2. For thicknesses less than 0.14 μm,
multiple scattering has not been attained and the curves are not

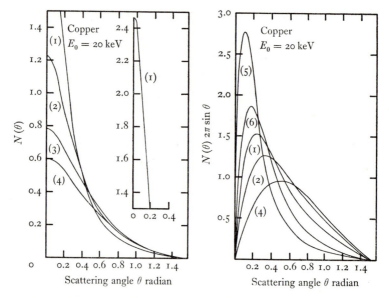

Fig. 4.2. The angular distribution of 20 keV electrons which have passed through films of copper of various thickness. After Cosslett and Thomas (1964b).

(1) $\rho x = 59\ \mu g/cm^2$ $x = 0.066\ \mu m$
(2) $\rho x = 80.4\ \mu g/cm^2$ $x = 0.090\ \mu m$
(3) $\rho x = 125\ \mu g/cm^2$ $x = 0.14\ \mu m$
(4) $\rho x = 172\ \mu g/cm^2$ $x = 0.193\ \mu m$
(5) $\rho x = 19.3\ \mu g/cm^2$ $x = 0.0216\ \mu m$
(6) $\rho x = 38.4\ \mu g/cm^2$ $x = 0.043\ \mu m$

Gaussian in form; scattering at small angles is accentuated because a considerable number of electrons pass through the film without collision.

Experiments with aluminium, silver and gold, as well as copper, indicated that the approach to multiple scattering is determined primarily by the number of elastic collisions, independently of the material and of the value of E_0. For all four materials at $E_0 = 10\ keV$ and $E_0 = 20\ keV$, multiple scattering is attained after 25 ± 5 collisions and the most probable angle of scattering is then $20° \pm 2°$. Assuming the number of elastic collisions to be given by (4.1), the curves of fig. 4.2 should apply equally well to other elements and to other values of E_0, so long as the film thicknesses are scaled in such a way as to keep $Z^{\frac{4}{3}}\rho x/A E_0$ constant.

Once the thickness has increased to the minimum value required

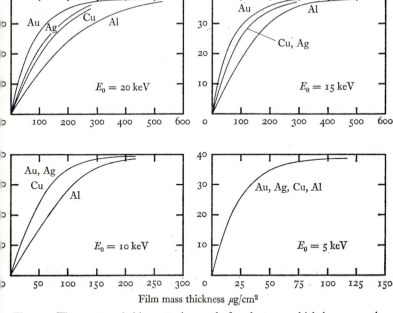

Fig. 4.3. The most probable scattering angle for electrons which have passed through various thicknesses of aluminium, copper, silver or gold. After Cosslett and Thomas (1964*b*).

for multiple scattering, further increase causes the most probable angle of deflection to rise to the asymptotic value characteristic of diffusion. For all four of the metals studied, this asymptotic value was $38°$, but the approach to it was not the same for the different materials and for the different values of E_0. The experimental results are summarized in fig. 4.3.

As a rough guide it may be said that diffusion conditions are attained after about 100 elastic collisions.

4.2. The range of electrons in a solid

4.2.1. Definitions of range

When a collimated beam of electrons, of initial energy E_0, falls normally on the surface of a solid, collisions with the atoms of the solid cause the electrons to change direction and to lose energy. It thus comes about that there is a certain depth from the surface beyond which the electrons do not penetrate and we can thus speak, in a general way, of the *range R* of an electron of energy E_0. If all of the electrons behaved in exactly the same way, R would be a definite quantity, but this is not the case and, in practice, we must adopt some arbitrary definition of what we mean by the range. At least four such definitions occur in the literature and these we now consider.

(a) The maximum range (R_{max})

If η_T is the fraction of the incident electrons which reach a depth of at least x below the surface, we define the maximum range R_{max} to be the minimum value of x which just reduces η_T to zero. In principle R_{max} can be found by measuring η_T for films of material of different thickness x, plotting η_T against x and extrapolating. Unfortunately, η_T tends to decrease exponentially with x, so the experimental value of the range is greatly affected by the sensitivity of the measuring instrument used to determine η_T. A somewhat better procedure is to plot η_T as a function of E_0 for a film of constant thickness, to give curves of the form shown in fig. 4.4 for copper and gold respectively. However, the sensitivity of the measuring instrument remains an important factor and values of R_{max} obtained by different observers show considerable scatter.

(b) The extrapolated range (R_x)

It will be observed that each of the curves in fig. 4.4 has a point of inflexion in its lower part and is therefore almost straight in this region. If, for a film of thickness x', this linear portion of the curve is extrapolated to cut the axis of E_0 at some value E_0', then x' is said to be the extrapolated range R_x for electrons of initial energy E_0'. The justification for this somewhat arbitrary procedure is that it leads to a definition of range which is independent of the sensitivity of the measuring instrument. From the shapes of the curves in fig. 4.4, it is clear that R_x will always be less than R_{max}.

(c) The Thomson–Whiddington range (R_{tw})

Instead of measuring the fraction of incident electrons transmitted through a film of given thickness, one may study the loss of energy

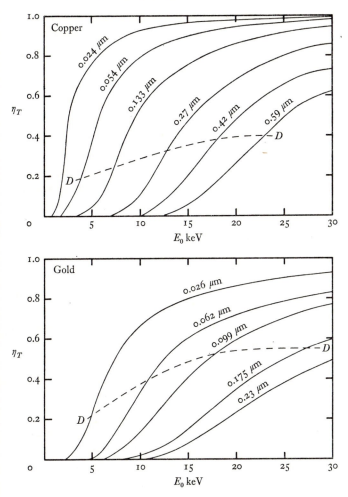

Fig. 4.4. Fractional transmission of electrons as a function of incident energy. After Cosslett and Thomas (1964c).

which they suffer. This was first done by Whiddington (1912), and the resulting Thomson–Whiddington law

$$E_0^2 - E_p^2 = bx, \qquad (4.6)$$

has been widely quoted. In this expression, E_p is the most probable value of the energy of the transmitted electrons and b is a constant for a given value of E_0.

Clearly we can take the range of the incident electrons of energy E_0 to be the minimum thickness of material which reduces the most probable energy of the transmitted electrons to zero. This is known as the Thomson–Whiddington range R_{tw} and (4.6) has been widely misinterpreted to indicate that

$$R_{tw} = kE_0^2,$$

where k is a constant. This is not the case, however, because b varies with E_0.

More recent work has shown that (4.6) is not valid for films whose thickness is less than about half of R_{tw}, so R_{tw} is not a particularly useful concept for our purpose and we shall not refer to it again.

(d) The mean range (R_m)

If the mean energy E_m of the transmitted electrons is measured, rather than their most probable energy E_p, a law of the Thomson–Whiddington type

$$E_0^2 - E_m^2 = b'x \qquad (4.7)$$

is found to hold quite accurately for a wide range of film thickness. We therefore define the mean range R_m, for incident energy E_0, to be the minimum thickness x that reduces E_m to zero. Once again, b' varies with E_0, so R_m is not proportional to E_0^2.

4.2.2. Values of range

Cosslett and Thomas (1964c) have measured the various ranges in aluminium, copper, silver and gold, for initial electron energies up to 20 keV. Their paper summarizes the results obtained by other experimenters, contains a comprehensive list of references and discusses in considerable detail the relations between the experimental measurements and the various theories which have been put forward to explain them. The results which follow are taken from this paper, but are presented in forms most likely to be of value to the user of a scanning electron microscope.

Numerical values of the ranges are given in table 4.3. Measurements of these quantities are not easy to make and the experimental error is unlikely to be less than ± 6 per cent. Thus, for aluminium and copper, any difference between R_x and R_m is barely significant. For silver, R_m is consistently slightly higher than R_x for the larger values of E_0 and this trend is still more marked in gold. In all cases R_{max} is, as we should expect, considerably greater than either R_x or R_m.

TABLE 4.3

E_0 (keV)	2.5	5	7.5	10	12.5	15	17.5	20	
R_{max} (μm)	0.19	0.45	0.74	1.15					
R_x (μm)	0.09	0.28	0.52	0.85	1.3	1.8	2.4	3.1	Aluminium
R_m (μm)	0.09	0.28	0.57	0.89	1.3	1.8	2.2	2.8	
R_{max} (μm)	0.084	0.17	0.28	0.44	0.64				
R_x (μm)	0.045	0.12	0.21	0.33	0.45	0.59	0.75	0.91	Copper
R_m (μm)	0.038	0.11	0.20	0.31	0.43	0.56	0.70	0.86	
R_{max} (μm)	0.076	0.14	0.24	0.38	0.53	0.72			
R_x (μm)	0.048	0.10	0.17	0.27	0.38	0.49	0.62	0.75	Silver
R_m (μm)	0.036	0.10	0.19	0.29	0.40	0.52	0.66	0.81	
R_{max} (μm)	0.039	0.078	0.14	0.23					
R_x (μm)	0.021	0.052	0.09	0.15	0.20	0.26	0.34	0.41	Gold
R_m (μm)	0.023	0.065	0.12	0.19	0.26	0.34	0.42	0.52	

For all four elements the relation between R_m and E_0 over the range 9–18 keV, can be represented by an expression of the form

$$R_m = pE_0^q, \qquad (4.8)$$

where the appropriate values of p and q are as shown.

		Al	Cu	Ag	Au
p		200	96	90	58
q		1.65	1.5	1.5	1.5

In the absence of comprehensive experimental data, estimates of ranges for other values of E_0 and in other elements can only be made by extrapolation of such measured values as are available. Bearing in mind that the range in any element is determined by numerous collisions between the electrons and the atoms, we should expect the range to vary in a regular manner with the atomic number Z and detailed theoretical treatments of the problem support this expectation. Furthermore, it transpires that the product ρR, where ρ is the density and R is the range (whether R_{max}, R_x, or R_m) varies relatively slowly with Z. This product is therefore a convenient

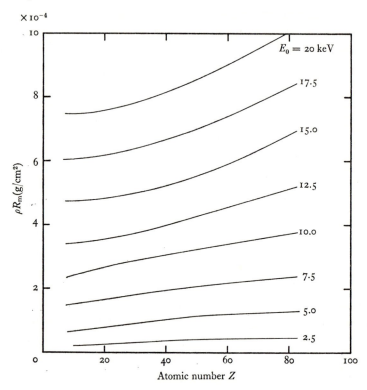

Fig. 4.5. Variation of ρR_m with atomic number Z.

quantity to use for purposes of extrapolation. In fig. 4.5 curves of ρR_m have been plotted against Z for various values of E_0; they should enable reasonably accurate estimates of R_m to be made for any element and for any value of E_0 in the range 2.5–20 keV.

We may sometimes wish to have information about the energy of electrons that have been transmitted through a film of thickness x which is less than the range, and we have already seen that their mean energy E_m is given by an expression of the form

$$E_0^2 - E_m^2 = b'x, \tag{4.9}$$

where b' is a constant for any given value of E_0 and for any given element. It transpires that b'/ρ does not vary very rapidly with atomic number and is therefore a convenient parameter to use for

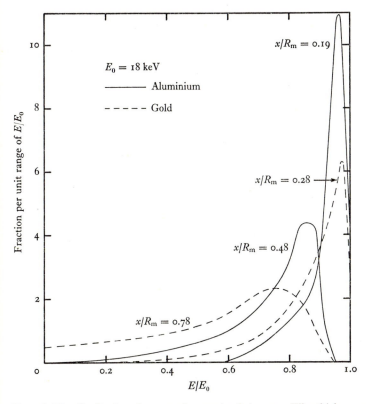

Fig. 4.6. The distribution of energy of transmitted electrons. Film thickness x is expressed as a fraction of the mean range R_m.

extrapolation. The experimental values of b'/ρ for aluminium, copper, silver and gold, listed in table 4.4, are taken from Cosslet and Thomas (1964c).

Cosslett and Thomas also give some measurements of the distribution of energies of the transmitted electrons and in fig. 4.6 we have replotted four of their curves, two for aluminium and two for gold, for $E_0 = 18$ keV. Film thicknesses are given as fractions of the mean range R_m and it will be seen that, even with two elements of such widely differing atomic numbers as aluminium and gold, the four curves form a fairly consistent set. We should thus expect curves for other elements and the same value of E_0 to fit into the same set. These curves refer to electrons incident normally on the film and leaving the film in this same direction. Such measurements as were made on electrons leaving in other directions, up to $55°$ from the normal, gave very similar energy distributions.

TABLE 4.4. *Values of b'/ρ when E_0 is in keV; ρ in g/cm^3; x in μm*

	E_0	9	15	18
	Al	40	46	52
b'/ρ	Cu	34	39	49
	Ag	30	38	46
	Au	27	31	38

4.3. Back-scattering of electrons

4.3.1. Introduction

From what has already been said it will be clear that, when a beam of electrons is incident normally on the surface of a solid, some of the electrons will be deflected through large angles. This may happen because an electron has passed near the nucleus of a scattering atom on one or more occasions, or because it has suffered a large number of small-angle deflections. When the total scattering angle exceeds $90°$, an electron may emerge through the surface by which it entered the solid, though it will not necessarily do so since it may be deflected again before reaching this surface. Electrons which do emerge in this way are said to be *back-scattered*: they are of obvious importance in any consideration of the operation of a scanning electron microscope.

When investigating this phenomenon experimentally, there is no means whereby back-scattered electrons can be distinguished from electrons produced by secondary emission, which we shall consider later. However, the former group tend to have relatively high energies, while most of the latter have much lower values. By a convention which has been widely adopted, electrons leaving the surface with energies greater than $50\,eV$ are taken to be back-scattered, while those with energies below $50\,eV$ are regarded as secondaries.

Much of the experimental work on back-scattering has been carried out to test the theories that have been proposed to explain the observed phenomena. With this aspect of the matter we shall not be concerned; we are interested only in those features which are of direct relevance to scanning electron microscopy. The general account which follows is based on the work of Palluel (1947), Sternglass (1954), Kulenkampff and Ruttiger (1954, 1958), Kulenkampff and Spyra (1954), Kanter (1957, 1964), Weinryb and Philibert (1964), Cosslett and Thomas (1965) and Bishop (1966).

4.3.2. Normal incidence on bulk solids

A quantity which has been measured by a number of observers is the ratio of the total back-scattered current (excluding secondaries) to the incident electron current falling normally on the surface; it is known as the back-scattering coefficient η_B. Over the range with which we are concerned, this quantity does not vary greatly with the energy of the incident electrons, but it does depend quite strongly on the nature of the scattering material. There is some evidence that it does not vary smoothly with the atomic number Z, but is influenced also by the extranuclear electronic structure. However, deviations from regular variation with Z are not very much larger than the experimental uncertainty and, for our purpose, are not important. In fig. 4.7 the coefficient is plotted against Z for incident energies E_0 of 5 keV and 30 keV.

For incident energies below 5 keV the back-scattering coefficient for elements of high atomic number tends to fall but this effect is much less marked for elements of low atomic number, whose coefficient is already small, even for higher values of E_0. At 0.5 keV, platinum, tungsten, molybdenum, silver and iron appear to have coefficients not very different from 0.2, while carbon retains the value of 0.1 which is also found at much higher values of E_0. Results obtained by different observers do not always agree very closely and the figures just given should be taken to indicate only the general trend of the coefficient as E_0 is reduced.

In addition to the total back-scattering coefficient η_B, we need to know how the emergent electrons are distributed in direction. If $\eta(\theta)\,d\Omega$ is the fraction of the incident electrons scattered into an element of solid angle $d\Omega$, making angle θ with their direction of incidence, to a close approximation

$$\eta(\theta) = (\eta/\pi)\,|\cos\theta|, \tag{4.10}$$

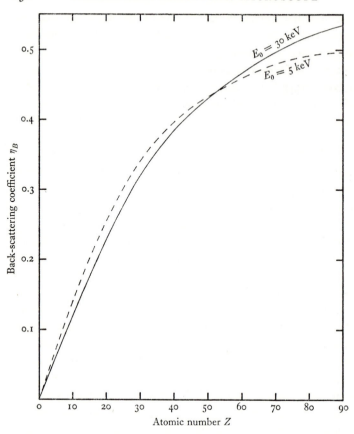

Fig. 4.7. Variation of the back-scattering coefficient with atomic number.

that is, the scattered electrons have very nearly a Lambert cosine distribution. This relation appears to hold accurately for $135° < \theta < 180°$. For values of θ between $90°$ and $135°$ the values of $\eta(\theta)$ are somewhat smaller than (4.10) would indicate but the deviation is of little importance for our purpose. The distribution in angle seems to be independent of the initial energy of the electrons, at any rate over the range 5 to 30 keV.

As we should expect, the back-scattering process involves loss of energy and typical curves showing the way in which the energies of the scattered electrons vary with the material and with the direction of scattering are reproduced in fig. 4.8 (Bishop, 1966). So far as

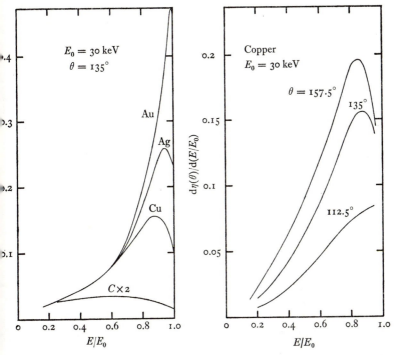

Fig. 4.8. Distribution of energies of back-scattered electrons. After Cosslett and Bishop (1966) in *4th Congress on X-ray optics and microanalysis* ed. Castaing, Deschamps and Philibert (Paris: Hermann).

scanning electron microscopy is concerned the important fact emerges that, except perhaps with elements of very low atomic number, the majority of the back-scattered electrons will have lost less than half of their initial energy. They will therefore be travelling with high velocities and their trajectories are unlikely to be greatly affected by electric or magnetic fields existing in the specimen chamber.

4.3.3. Normal incidence on thin films of material

In the scanning electron microscope it may often happen that the specimen under observation has a structure containing thin films of material, or has been coated with a thin film of metal to prevent charging under the action of the electron probe. It is therefore of interest to have information about back-scattering from thin films.

Cosslett and Thomas (1965) give results of measurements on back-scattering from copper and gold films and their data are used in the following discussion. As the thickness x of a film is increased, the back-scattering coefficient η will rise towards the value η_B which is obtained with bulk material. On general grounds we should expect this asymptotic value to be reached when x is roughly equal to half the mean range R_m of electrons in the material, since an incident electron which penetrated bulk material to a depth greater than $\frac{1}{2}R_m$ would have to travel a total distance greater than R_m to be back-scattered. This general prediction is in agreement with the experimental results.

The above fact suggests that it might be instructive to present the experimental results in such a way that the ratio η/η_B of the back-scattering coefficient for a film of thickness x to the coefficient for bulk material, is plotted against x/R_m. When this is done, it is found that the results given by Cosslett and Thomas for copper and for gold respectively, over a range of initial energies from 5 keV to 30 keV, all lie quite close to a single curve which is shown in fig. 4.9. In fact the deviation of any one of their curves from fig. 4.9 is probably not much greater than the experimental error.

It is not suggested that this method of plotting the results necessarily has physical significance; in fact there is reason to believe that the mechanisms leading to back-scattering (single large-angle deflections and multiple small-angle deflections respectively) have a relative importance which depends on the atomic number. However, the fact that fig. 4.9 holds good for two elements of such different atomic number as copper and gold, indicates that it is unlikely to be too greatly in error if applied to other elements.

Kanter (1964) has investigated the angular distribution of electrons back-scattered from films of gold ($x = 0.01\,\mu$m) and aluminium ($x = 0.02\,\mu$m) deposited respectively on films of Al_2O_3 ($x = 0.02\,\mu$m). For the gold film he found that the circular polar diagram characteristic of a cosine distribution became progressively flattened as the initial energy was increased and developed incipient side-lobes. At $E_0 = 10$ keV these side-lobes had maxima at about $45°$. With the aluminium film this effect was not observed and the cosine distribution persisted.

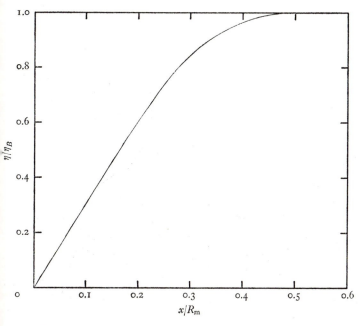

Fig. 4.9. Variation of back-scattering coefficient with film thickness.

η = coefficient for a film of thickness x.
η_B = coefficient for bulk material.
R_m = mean range of electrons.

4.3.4. Bulk material; oblique incidence

In the scanning microscope it will rarely happen that the electron probe strikes the specimen normally. Even if the average surface of the latter is at right angles to the probe, there will often be local irregularities which cause the angle of incidence to differ from zero. It is therefore of importance to consider how back-scattering is affected by oblique incidence.

As the angle of incidence increases from zero, the paths of the electrons in the material will, on the average, lie nearer to the surface. We should therefore expect the total back-scattering coefficient to increase, tending towards unity as the incident angle γ approaches 90°. We should also expect the scattered electrons to leave the surface preferentially in the direction of incidence.

Both of these expectations are justified, as is shown by some measurements reported by Kanter (1957). Fig. 4.10, taken from his results, refers to back-scattering in the plane containing the incident probe and the normal to the surface. It shows that the scattered electrons do indeed leave the surface preferentially in the direction of the incident beam, but that the effect does not become very marked until the angle of incidence exceeds 55°. Moreover, it is only in this particular plane that the effect is very noticeable, so that the total back-scattering coefficient is not increased very greatly by

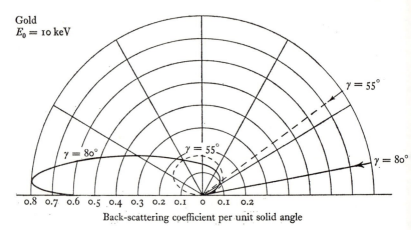

Back-scattering coefficient per unit solid angle

Fig. 4.10. Variation of back-scattering coefficient with angle of incidence. After Kanter (1957) *Ann. d. Physik*, **20**, 144.

oblique incidence. For polished gold it rises from 0.5 to 0.6 between $\gamma = 0°$ and $\gamma = 45°$. For polished aluminium the corresponding rise is from 0.15 to 0.3. In both cases the change is less if the metals have matt surfaces.

4.4. Distribution of charge and energy

When an electron beam enters a material, some of the electrons will be neither transmitted nor back-scattered. These represent loss of charge to the material which, if the latter is not an insulator, will be conducted away through the connecting lead. When the specimen is an insulator it becomes negatively charged, but the extent to which this happens is usually determined by secondary emission; it will be considered later (§ 4.6).

The absorption of incident electrons also represents loss of energy, most of which appears as heat in the material. In this case secondary emisson plays little part, since the secondary electrons are of low velocity and carry away very little energy. We may sometimes wish to know the energy dissipated in a thin film of material when a beam of electrons is incident on it, but the information needed to enable a detailed calculation to be made is rarely available· However, a rough estimate may be made as follows.

For normally incident electrons of energy E_0 keV and incident current I amperes, the incident power is $E_0 I$ kW. From the curves of fig. 4.5, we can estimate the mean range R_m. By the definition of R_m this quantity is equal to the value of x obtained by putting E_m equal to zero in (4.9). Thus we obtain the value of b' in (4.9) corresponding to the given value of E_0, and this enables us to calculate the mean energy E_m of electrons which have passed through a film of thickness x, which is less than R_m. From the curves of fig. 4.1 we estimate the current I' which succeeds in penetrating the given film and we may then say that the power removed by this current is $I'E_m$.

We next consider the back-scattered electrons and, from fig. 4.7, estimate the back-scattering coefficient η_B for bulk material. The curve of fig. 4.9 then allows us to deduce the corresponding coefficient for the given film, and hence the current I'' which is back-scattered from it. Relatively little information is available on the mean energy of these electrons, but if we take it to be $0.8E_0$ we are unlikely to be very greatly in error. For elements of low atomic number the loss will be greater than we have suggested, but the back-scattering coefficient for these elements is low, so the overall power balance will only be slightly affected. Thus, for the power W dissipated in the film, we have

$$W = IE_0 - I'E_m - 0.8I''E_0. \qquad (4.11)$$

If the electron beam is incident normally on bulk material, only back-scattering need be considered and the total power W' dissipated in the material is

$$W' = IE_0 - 0.8I''E_0, \qquad (4.12)$$

where I'' is now calculated from the back-scattering coefficient for bulk material. It may be of interest to know how the dissipation of **this power, per unit thickness** of material varies with the depth below the surface. At first sight, it might be thought that the power dissipated in a top layer of thickness x would be the same as that dissipated in a film of thickness x, as calculated above, but this is not the case. Some electrons which pass through this top layer of the bulk material will be back-scattered at greater depths and will return to the top layer to give up their energy there. This problem has been considered by Cosslett and Thomas (1965) and the curves

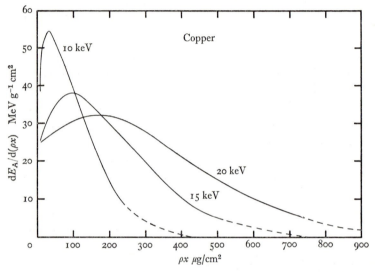

Fig. 4.11. Dissipation of electron energy as a function of depth below the surface. After Cosslett and Thomas (1965).

for copper given in fig. 4.11 are based on their paper. For each curve the experimental results are extrapolated (broken lines) to cut the axis at the maximum range R_{max}. Curves for gold, in terms of mass thickness ρx, were of the same general form with similar absolute values, but the peaks, particularly for the higher values of initial energy, were sharper and occurred at lower values of ρx.

4.5. Secondary electron emission

4.5.1. Introduction

In the preceding sections we have been concerned with the subsequent behaviour of a beam of electrons when it strikes solid material, and we have seen that this behaviour is largely determined by elastic collisions with the atoms of the material. However, some of the collisions are inelastic and these may cause secondary electrons to be liberated from the atoms. These secondaries move off in all directions and some of them will be directed towards the surface on which the initial beam was incident. If, when they reach the surface they have sufficient energy to overcome the potential barrier characteristic of the work function of the material, they will be emitted and it is to these electrons that we give the name *secondary emission*.

There is no experimental method by which we can distinguish between an electron that has been emitted in this way and a primary electron that has been back-scattered. However, we have seen that most of the back-scattered electrons retain a high proportion of their initial energy, while it will shortly appear that most of the secondaries have energies less than a few tens of electronvolts. It is therefore a widely accepted convention to regard all electrons with energies greater than 50 eV as back-scattered primaries and all those with energies less than 50 eV as true secondaries. In general, the fraction of the total which have energies in the vicinity of 50 eV is very small.

Secondary emission has been studied since the beginning of the century and the literature devoted to it is very considerable. We shall rely largely on reviews that have been published by Bruining (1954), by Kollath (1956) and by Hachenberg and Brauer (1959) each of which contains an extensive bibliography. A high proportion of the work that has been done is not very relevant to scanning electron microscopy, partly because measurements have often been restricted to initial energies of the incident electrons which are much lower than those commonly employed in the scanning microscope. In addition, much of the research has been devoted to theoretical studies of the phenomena and to experimental tests of the various hypotheses that have been put forward. At the present time, it cannot be said that a really satisfactory theory of secondary emission is available. We have a good general picture of the various factors that govern the phenomena, but it is not yet possible to make detailed calculations with any certainty.

For our present purpose we need to know how the generation of secondary electrons depends on the material and on the energy and direction of the incident beam. We also need information about the distribution of directions and of energies of the secondary electrons. In the following sections we give a general account of the experimental evidence relating to these matters; we shall not be concerned with theoretical explanations of the phenomena.

4.5.2. The secondary-emission coefficient when the primaries are incident normally

In all cases the number of secondary electrons generated is proportional to the number of primaries. We therefore define a secondary-emission coefficient δ by the relation

$$\delta = \frac{\text{number of secondaries (with energy less than 50 eV)}}{\text{number of primaries}}.$$

When comparison with theory is to be made, it is usual to subtract from the number of primaries the elastically back-scattered electrons, since these latter have produced no secondaries. This refinement is of no importance for our purpose and we shall not refer to it again.

From quite simple arguments we can see how δ is likely to vary with the energy E_0 of the primary stream. Since a secondary electron, to be emitted, must acquire sufficient energy to surmount the surface potential barrier, δ will be zero until E_0 is at least a few electronvolts. As E_0 increases we should expect δ to rise, but two opposing effects now come into play. An increase in E_0 certainly leads to the generation of a greater number of secondaries in the material, but it also gives the primaries greater penetrating power so that the secondaries are generated at greater depth. As they travel towards the surface they are liable to absorption, so many of them are never emitted. It thus comes about that as E_0 is increased, δ rises to a maximum value $\delta_{\max}$, after which the penetration effect becomes dominant, and δ decreases steadily with further increase in E_0.

It is found that, if $\delta/\delta_{\max}$ is plotted against $E_0/E_{0,\max}$, where $E_{0,\max}$ is the value of E_0 at which $\delta_{\max}$ occurs, a single normalized curve represents quite closely the behaviour of a considerable number of metals. An average curve of this kind is reproduced in fig. 4.12 for values of $E_0/E_{0,\max}$ up to four. The much higher values of $E_0/E_{0,\max}$ that are of particular interest in scanning microscopy, will be considered later. For semiconductors and insulators the normalized curve is of roughly the same shape as for metals, but different materials can no longer be represented as accurately by a single curve.

Turning now to absolute magnitudes, a representative list of values of δ_{max} and $E_{0,\,max}$ is given in table 4.5. A more comprehensive list would serve little purpose, since the values depend greatly

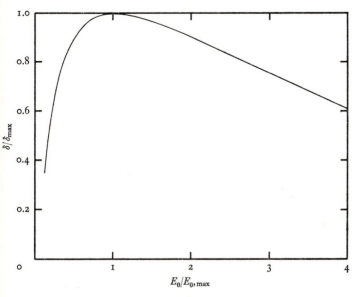

Fig. 4.12. Generalized curve of secondary-emission coefficient for metals. After Kollath (1956). In *Hand. d. Phys.* ed. Flügge, **21**, 232 (Berlin: Springer).

on the physical state of the surface of the material, on layers of adsorbed gas and so forth. For a specimen in the scanning microscope we are unlikely to have detailed information on these factors and all we need is a general idea of the trend of the values.

For almost all metals δ_{max} is in the region of unity, being rather less than unity for the alkalis and alkaline earths and rather greater than unity for most of the others. Similarly $E_{0,\,max}$ ranges from about 200 to 800 V, the smaller values being found for metals with low values of δ_{max}. Semiconducting elements and compounds are quite similar to metals but insulators generally have considerably higher values of δ_{max}, usually at somewhat higher values of $E_{0,\,max}$.

The value of δ_{max} for any material depends partly on the number of secondary electrons generated within the material and partly on the fraction of these which escape through the surface. It seems likely, on theoretical grounds, that it is a difference in the second of these factors which causes δ_{max} to be higher for insulators than for metals; absorption of secondary electrons within the material is weaker in insulators.

TABLE 4.5

	Material	δ_{max}	$E_{0, max}$ volts
Metals	Li	0.5	85
	Mg	0.95	300
	Al	0.95	300
	Cu	1.3	600
	W	1.35	650
	Ta	1.1	800
Semiconductors	Si (single crystal)	1.1	250
	C (diamond)	2.8	750
	C (graphite)	1	250
	PbS	1.2	500
Insulators	Mica	2.4	300–384
	SiO_2 (quartz)	2.4	400
	Al_2O_3 (layer)	1.5–9	350–1300
	NaCl (single crystal)	14	1200
	NaCl (evapd.-layer)	6–6.8	600
	MgO (single crystal)	23	1200

The curve of fig. 4.12 gives a good indication of the way in which δ varies with E_0 for values of E_0 up to a few keV. For purposes of scanning microscopy we should like to know δ for E_0 up to at least 30 keV but, unfortunately, very few measurements appear to have been made in this range. There is general agreement that δ continues to fall steadily as E_0 is increased and, to an ever greater extent, the true secondary emission becomes overshadowed by back-scattered electrons.

The back-scattered electrons themselves, on their way out of the material, make inelastic collisions and thus produce secondaries. This effect has been considered by Seiler (1967), who concludes that, because of their lower velocities and greater inclination to the normal, the back-scattered electrons are from four to six times as efficient in producing secondaries as are the primary electrons. He estimates that the fraction of slow secondaries generated by the back-scattered electrons rises from 0.25 when the back-scattering coefficient η_B is 0.05, to 0.7 when η_B is 0.45. Since η_B depends quite strongly on atomic number (fig. 4.5), it is not to be expected that the variation of δ with E_0 could be represented, over a wide range, by a single curve for all materials, as was the case for very low values of E_0 where back-scattering is much less important. The effect under consideration will compensate the falling off of δ with increasing E_0 to a greater extent in elements of high atomic number than those of low atomic number. This prediction seems to be in agreement with the experimental evidence.

From the very few measurements that are available, we may perhaps conclude that, for materials containing atoms of high atomic number, δ is likely to fall to 0.5 δ_{max} somewhere in the region of $E_0 = 8\,\text{keV}$, and to 0.25 δ_{max} when E_0 reaches 30 keV. When elements of low atomic number are involved, the fall should be rather more rapid, δ reaching perhaps 0.15 δ_{max} at $E_0 = 30\,\text{keV}$. Until further measurements have been made, these rough estimates should be treated with considerable reserve.

4.5.3. Influence of the angle of incidence

The foregoing results relate to an incident electron beam impinging normally on the material surface, but this condition will rarely be met in a scanning microscope. Several investigators have shown that, when the incidence is oblique, larger values of δ are obtained, and this can be explained from the following considerations.

Secondary electrons are generated all along the path of a primary electron but, as they move through the material, many are effectively absorbed, as a result of loss of energy or otherwise. Many of the experimental facts are consistent with the hypothesis that, after travelling a distance x through the material, the number of second-aries has decreased by a factor $\exp(-\alpha x)$, where the absorption coefficient α has a value of the order of $10^6\,\text{cm}^{-1}$. It thus comes about that almost all of the observed secondary emission results from electrons that were generated within a distance of the order of 0.01 μm from the surface. Within this distance and for reasonably high values of E_0, the primary electrons will have lost little of their initial energy and many of them will have been deflected through only small angles from their initial direction. With these assumptions, we take the number of secondaries generated in any element of the primary path to be proportional to the length of that element. If the primary beam is incident normally, the length of its path lying between depths x and $x + dx$ is, of course, dx. For incidence at angle θ, however, the length becomes $dx \sec \theta$. This is true for every element of path which is sufficiently near the surface to contribute to the observed secondary emission, so we may expect that

$$\delta_\theta = \delta_0 \sec \theta. \qquad (4.13)$$

This result is roughly in accord with experiment.

Bruining (1954) treats the problem somewhat differently and arrives at the expression

$$\ln (\delta_\theta / \delta_0) = \alpha x_{\mathrm{m}} (1 - \cos \theta), \qquad (4.14)$$

where x_{m} is the 'mean depth of origin' of the secondaries. This also is in quite good agreement with experiment and, using for α a value for nickel given by Becker (1929), viz $\alpha = 1.5 \times 10^6 \, \mathrm{cm}^{-1}$, he finds x_{m} to be approximately equal to 0.003 μm. This is consistent with measurements that have been made on secondary emission from thin films at normal incidence. For metals, it is found that δ_0 reaches a limiting value when the thickness of the film exceeds about 0.01 μm. This value would be the maximum depth from which secondaries pass through the surface; the mean depth would be less. For a film of KCl, a maximum depth of 0.05 μm was found. This confirms the suggestion made earlier that α is less in insulators than in metals. The above discussion ignores the production of secondaries by back-scattered primaries. However, we should expect this effect to become relatively less important as θ increases.

4.5.4. Angular distribution of secondary electrons

If the secondary electrons just below the surface of the material were moving in random directions, we should expect the emitted electrons to have a distribution of directions in accordance with Lambert's cosine law. Experiment shows that this law is, in fact, obeyed very closely. Small deviations may occur at values of E_0 of a few tens of electronvolts, but these are of no importance for our purpose. The distribution does not seem to be affected by the crystal structure of the material and is practically independent of the angle of incidence of the primary beam. If the total number of emitted secondary electrons is N and if $N(\theta)\,d\Omega$ is the number emitted into an element of solid angle $d\Omega$ which makes angle θ with the normal, we may write

$$N(\theta) = (N/\pi)\cos\theta. \qquad (4.15)$$

4.5.5. Energy distribution of secondary electrons

When the energy of the incident beam exceeds, say, 100 eV, the energy spectrum of the emitted secondary electrons exhibits a pronounced maximum. For a number of metals the distribution of

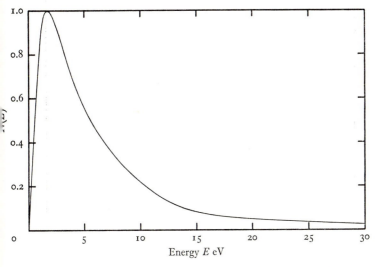

Fig. 4.13. Average distribution of energies of secondary electrons from metals.

energies is represented quite closely by the average curve drawn in fig. 4.13. The position of the maximum may vary a little from one metal to another but generally lies in the range 1.3 to 2.5 eV. The distribution does not seem to depend to any great extent on the direction in which the electrons are emitted. Semiconductors have distribution curves similar to those for metals.

With insulators the distribution curve is of the same form as with metals, but there is greater variation from one material to another and from one value of E_0 to another. In general the maxima tend to be sharper than those exhibited by metals, particularly at high values of E_0.

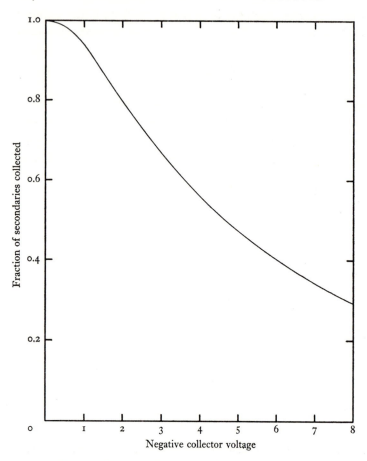

Fig. 4.14. Fraction of secondary electrons from metals which
are collected, as a function of negative collector voltage.

In scanning electron microscopy, we may sometimes wish to
investigate differences of potential on the surface of the specimen
by maintaining the collector at a small negative potential with
respect to the specimen. With this arrangement, more secondaries
will be collected from areas at low positive potential than from those
at high potential, even though the secondary emission coefficient is
the same for both areas. It is therefore of interest to know how the
fraction of secondaries accepted by the collector depends on the
effective negative bias of the collector with respect to the specimen,

after correction for any contact P.D. that may exist. This relation can be obtained by integration of the area under the curve in fig. 4.13, and is plotted in fig. 4.14.

4.5.6. The time constant of secondary emission

After a beam of primary electrons strikes a material, time must elapse before the secondary emission reaches its final value and the build-up process can be characterized by a time constant. It has not yet proved possible to measure this time constant but experiment indicates that it is less than 10^{-10} s. It is therefore unlikely to cause trouble in the scanning electron microscope.

4.6. Electrical charging of the specimen

When the specimen under observation in a scanning microscope is a conductor, the charge conveyed to it by the incident electron beam will be carried away by the leads and the specimen can be maintained at any desired potential. When the specimen is an insulator, it is common practice to coat its surface with an evaporated film of a suitable metal and to make electrical connection between this film and the specimen stage. This prevents charging and the thickness of film used will depend on circumstances: it will commonly lie between 0.01 and 0.1 μm.

It may sometimes be desirable to examine an insulating specimen without coating it with a metal film and we therefore consider the charging process in rather more detail. We have already seen that, for most insulators, the maximum secondary emission coefficient δ_{max} is greater than unity and that δ varies with the energy E_0 of the incident electrons. If δ' is a modified coefficient which includes the back-scattered primaries as well as the true secondaries, the general way in which δ' varies with E_0 is as shown in fig. 4.15. In particular, there are two values of E_0, indicated by E_0^{I} and E_0^{II} respectively, for which δ' is equal to unity. If the specimen could be kept at a potential corresponding to either of these two values of E_0, there would be no net acquisition of charge. The value of E_0^{I} is usually a few tens of electronvolts, while E_0^{II} commonly lies in the range 1 to 10 keV.

Let us suppose that the specimen in a scanning microscope is initially at the potential of its surroundings and that an incident electron beam of energy E_0 impinges on it. If E_0 is less than E_0^{I}, the specimen will receive more electrons than it loses and will charge negatively. This will produce deceleration of the incident electrons, causing them to strike the specimen with energy less than E_0, leading to a lower value of δ'. The process is clearly cumulative and will go on until the energy of the incident electrons has been reduced to zero: operation under these conditions is impossible.

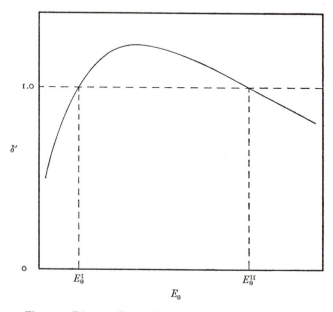

Fig. 4.15. Diagram illustrating the first and second crossover points E_0^{I} and E_0^{II}.

By a similar argument, if E_0 lies between E_0^{I} and E_0^{II}, the specimen will begin to charge positively. However, this process is rapidly brought to a halt because, as we have seen, most of the secondary electrons have energies of only a few electron volts. Thus, as soon as the potential of the specimen has become a few volts positive with respect to its surroundings, the slower secondaries will be unable to leave it and will re-enter the material. Equilibrium will be reached at a potential such that the total current leaving the specimen is exactly equal to the incident current.

Finally, if E_0 is greater than E_0^{II}, the specimen will charge negatively and the incident electrons will be decelerated. The process will go on until the energy of the incident electrons has been reduced to E_0^{II}, when equilibrium will be attained.

For the purpose of scanning microscopy it is important that E_0 should lie between E_0^{I} and E_0^{II} and that, as is normally the case, the potential of the specimen chamber with respect to the cathode should be that corresponding to E_0. Charging of the specimen will then change its potential by only a few volts and the energy of the incident electrons will be determined almost entirely by the potential difference between cathode and specimen chamber. Nothing can be gained by raising E_0 initially above E_0^{II}, since charging will, in any case, reduce the incident energy to E_0^{II}. Moreover, E_0^{II} depends on the properties of the material and is likely to vary as the incident probe is scanned over the surface of the specimen. Different areas of the specimen will be charged to different potentials and the resulting transverse electric fields will interfere with the collection of electrons. The operation of the microscope can hardly be satisfactory under these conditions.

CHAPTER 5

CONTRAST AND RESOLUTION

5.1. Contrast

5.1.1. General considerations

In this chapter we discuss factors which affect contrast and resolution in a scanning electron microscope, in which the display signal is derived from secondary electrons or electrons back-scattered from an opaque specimen. The type of scanning instrument in which the incident probe passes through a thin transparent specimen is thus specifically excluded, since quite different conditions obtain in this case.

The utility of a scanning microscope, for the observation of opaque specimens, depends to a very great extent on its ability to produce a 'three-dimensional picture' of the surface of the specimen and, in this connection, two very general points should always be kept in mind. The first is the very obvious one that, unless stereoscopic techniques are used, any 'three-dimensional picture', whether produced by photography, television, or a scanning microscope, is an optical illusion that depends for its effectiveness on the previous experience of the eye and brain of the observer. Information about distance at right angles to the plane of the picture is not contained in the picture itself and the observer deduces this information by interpreting lights and shadows on the basis of his previous visual experience. His deductions are by no means infallible; it is not uncommon for an area of the picture which appears to represent a depression in the specimen surface, to have the opposite appearance if the picture is turned upside down.

The second point concerns the process by which the image on the display tube of a scanning microscope is built up. Here there is an obvious analogy with the techniques of television, but there are also important differences. In a television system the initial signal is derived from light reflected by the object being televised; the image focused on the camera tube is an optical image which does not differ very greatly from that which would be focused on the retina of an observer looking directly at the object. Thus the complete television system is reproducing an initial visual image. In the scanning microscope the situation is different; the brightness of each point of the image depends on the number of electrons collected from the corresponding point of the specimen and it is clear that the efficiency of any area of specimen as a secondary emitter or backscatterer bears little relation to its ability to reflect light. Thus the ratio of brightness of two areas of the microscope image must usually be quite different from the ratio that would be obtained if the corresponding areas of the object could be observed visually by reflected light. Nevertheless, the image produced by the scanning microscope when interpreted sub-consciously by the eye and brain of the observer on the basis of prior visual experience, conveys the kind of picture that he would expect to obtain if direct visual examination of the specimen were possible. Whenever the microscope is used to examine a specimen whose surface topography is known from other evidence, our instinctive interpretation of the picture produced by the microscope coincides in detail with the specimen structure that is known to exist. It is pertinent to enquire why this should be the case.

5.1.2. Contrast and its interpretation

The generation of contrast in a scanning microscope and its interpretation by the eye of the observer are subjects on which very few experimental results have been published. The subject is a complex one, since several quite independent factors combine to produce the overall result, and further work is needed to establish their relative importance. In the discussion which follows, we draw attention to these various factors and attempt to assess their influence. We wish to emphasize, however, that our treatment must be regarded as a tentative one; there is much still to be learned about the mechanism of contrast.

The eye is constructed to appreciate an enormously wide range of brightness in any scene at which it is looking and, to enable it to do this, the 'sensation' which it passes to the brain is a very non-linear function of the intensity of light falling upon the pupil. Leaving aside the very real difficulties in measuring visual sensation, the following experimental result is relevant to our discussion.

If I and $I \pm \Delta I$ are two intensities of illumination which the eye is just able to distinguish when they occur in adjacent areas then, over the range of intensity with which we are concerned, $I/\Delta I$ is roughly proportional to the logarithm of I. Thus the ability of the eye to detect fractional changes of intensity between adjacent areas of a picture does not depend very markedly on their average intensity.

We turn now to the factors which can cause variations in intensity in the image produced by a scanning microscope. These can be classified as follows:

(a) Factors which depend directly on the local angle of incidence of the electron probe.

(b) Factors which depend on the nature of the material of the specimen.

(c) Factors which cause variation in the fraction of secondary and back-scattered electrons that are collected by the detector.

We shall later be examining each of these classes in detail and it will appear that there is some overlapping, in that some of the factors in (c) depend indirectly on the local angle of incidence of the probe. We shall see that (a) and (c), taken together, are largely determined by topographical features of the specimen surface; by hills, depressions, crevasses, holes and edges. But these are precisely those features that would cause changes in the intensity of reflected light if the specimen could be examined visually. The relative intensities of different areas of the image would not be the same in the two cases but, as we have seen, the human eye is more sensitive to changes in intensity than to absolute levels. Thus it is not surprising that the microscope image conveys to the brain of the observer almost exactly the same impression of the topological features of the specimen surface as that which he would expect to receive if he could examine the specimen visually by reflected light.

The case is quite different for the factors included in class (b). Changes of material may, of course, coincide with topographical features, but they need not necessarily do so. Thus the factors in class (b) cause contrast which is not of primary importance in telling us about the topographical appearance of the specimen. It conveys additional information about changes in the material of which the specimen is composed and is somewhat analogous to colour in a visual image. Pushing this analogy a little further, we may say that the factors included in classes (*a*) and (*c*) are those which provide us with a black-and-white image and, from photographic and television experience, we know that such an image gives very complete information about topographical features of the surface of an object.

We shall now show that, of the factors included in classes (*a*) and (*c*), those in class (*a*) are of primary importance in determining the general appearance of the image in a scanning microscope. We have seen that the image has a pronounced 'three-dimensional' appearance and this implies that the object has been 'viewed' from a closely defined direction. In the case of visual observation, the viewpoint is determined by the observer's eye; from each surface element of the specimen, only those light rays which leave in a particular direction can enter the eye and thus contribute to the image. There is nothing comparable with this in the electron collection system of a scanning microscope. The detector often subtends a large solid angle at the specimen and slow electrons leaving the specimen travel in curved trajectories under the influence of the electric field maintained between specimen and detector. Moreover, it is known that satisfactory images of certain parts of the specimen can often be obtained under conditions such that electrons from these parts could not have reached the detector without subsequent reflection from other surfaces. Thus the system by which electrons are collected is quite incapable of defining a direction from which the specimen is viewed. The factors in class (*c*) cannot be the primary ones in determining the general appearance of the image, except where they depend indirectly on the local angle of incidence of the probe.

Turning now to the factors in class (*a*), we again find a striking difference from the conditions which obtain in visual observation, where the object is generally illuminated by light reaching it from a variety of directions. In the scanning microscope, on the other hand, the incident electrons reach the specimen from a very sharply defined direction and this fact provides the 'viewpoint' that was missing from the collection system. The image which is finally displayed is the picture of the specimen as viewed along the incident probe. The general appearance of the image is determined primarily by the factors in class (*a*), with class (*c*) factors modifying the contrast between one topographical feature and another. Factors in class (*b*) are, to a great extent, independent of topography; they depend on changes in the material of the specimen.

The dominant part played by factors in class (*a*) is responsible for one of the most valuable properties of the scanning microscope; its ability to reveal wall structure inside relatively deep holes and crevasses in a specimen. The very narrow incident probe provides strong 'illumination' of such areas and the fact that the secondary and back-scattered electrons may reach the detector by roundabout paths, possibly after several reflections, in no way detracts from the quality of the final image. Comparison with the corresponding situation in an optical microscope emphasizes the great advantage of the scanning electron instrument for this particular application. This advantage is enhanced by the much greater depth of field of the scanning microscope.

5.1.3. Factors in class (*a*)

Sometimes the specimen is mounted so that its mean surface is normal to the incident electron probe, but more often, to facilitate the collection of electrons leaving the specimen, the mean surface is at an angle to the probe. In either event, local irregularities in the surface are likely to cause the local angle of incidence to vary from point to point and it is the effect of this variation that we must now consider.

Let θ be the local angle of incidence, S the cross-sectional area of the probe and let n be the number of electrons delivered by the probe in unit time. These n electrons fall on an area $S \sec \theta$ of the specimen, so the number of electrons which the specimen receives per unit area per unit time can vary markedly from point to point. It might be thought that this variation would result in contrast in the image, but this is not the case. The image of an area of the specimen for which the probe angle of incidence is θ is foreshortened by a factor $\sec \theta$, so the reduction in the density of electrons falling on the specimen is exactly compensated by foreshortening in the image. The number of incident electrons per picture point of the image remains constant. A somewhat related effect concerns shadowing of the specimen surface. A protruberance in one area of the surface may make it impossible for incident electrons to reach another area, so no signal can be received from this second area. Will this not cause contrast by making the second area appear dark in the image? The answer to this question is that the second area will not appear in the image. The view of the object that is being imaged is the view seen when looking along the incident beam; we must no more expect to see the second area than we should expect to see the back of a house in a photograph taken from the front.

If the specimen surface is heavily indented, electrons back-scattered from one area may strike a second area and produce secondaries which are then attracted to the detector. The total time for such an occurrence would normally be small compared with that required for the incident probe to move from one picture point to the next, so all of the electrons reaching the detector, from which-ever of the two areas, will contribute solely to the brightness of the first area. This brightness may well be different from what it would have been if the second area had not intercepted some of the back-scattered electrons, but it is not possible to make any useful general statements about the effect. Rather similar effects can occur when protruberances on the specimen have the form of thin plates or rods of small diameter. Some of the incident electrons may then generate a normal signal at the surface of the protruberance, while others pass through it to generate an additional signal from an area which one might expect to be shadowed by the protruberance. When this occurs, 'ghost' details may appear in the image on the display tube but, once again, it is not helpful to attempt a general treatment of the phenomenon.

We turn next to a much more straightforward effect which is susceptible to general treatment (Everhart, Wells and Oatley, 1959). We have already seen (§4.5.3) that, when the incident probe strikes the specimen surface at an incident angle θ, the number of secondary electrons generated is roughly proportional to $\sec\theta$, and that this result is largely independent of the material of the specimen. Assuming, for the moment, that the brightness B in the image is proportional to the number of secondary electrons leaving any area of the specimen, the change ΔB in B which results from a change $\Delta\theta$ is given by

$$\Delta B/B = \tan\theta . \Delta\theta. \tag{5.1}$$

Thus, for $\theta = 45°$ a change $\Delta\theta$ of one degree would produce a change of B about 1.75 per cent, which should be detectable in the image. It therefore appears that we have in this effect a mechanism which can account for the observed contrast between adjacent areas of a specimen, for which the incident angle θ differs by quite a small amount.

In practice, the detector will receive some of the back-scattered electrons, as well as secondaries, and the total number of back-scattered electrons is largely independent of the angle of incidence of the probe (§4.34). For this reason, (5.1) may overestimate the change of brightness to a small extent. Indirectly, the angle of incidence may affect the number of back-scattered electrons entering the detector (§4.34), but this effect is unlikely to be of great importance.

5.1.4. Factors in class (*b*)

Dealing first with secondary emission, we have seen (§4.5.2) that, for metals and semiconductors, the secondary-emission coefficient does not usually differ very greatly from unity. For these substances, variation of the coefficient from one material to another in the specimen may well produce slight contrast, but we should not expect the effect to be very marked. When we turn to insulators, much higher coefficients are often found and these would be expected to give rise to strong contrast. However, this effect is often masked by the fact that the insulator must be coated with a thin conducting layer to prevent charging of the specimen. The secondary-emission coefficient may then be partly characteristic of the coating rather than of the underlying insulator.

A special case arises when a conducting specimen has areas which are coated with a very thin layer of an insulator, for example, an oxide or sulphide of a metal. In such a layer, the incident probe will generate hole/electron pairs to produce 'bombardment conductivity' which may well prevent charging of the specimen. The large difference of secondary-emission coefficient between the insulating layers and other areas of the specimen may then result in very strong contrast. As a rule, this effect is important only when the insulating layer is very thin. Some of the holes and/or electrons produced by the primary probe enter traps in the material and, if the thickness is too great, the resulting space charge reduces the conductivity to such an extent that charging of the specimen becomes troublesome.

When charging is severe, the specimen must be given a conductive coating, usually by evaporating on to its surface a metallic film whose thickness is of the order of 0.01 μm, though anti-static preparations are sometimes used, particularly with synthetic fibres. The nature of any such coating may profoundly affect both the secondary and the back-scattered electrons. Of the metals, aluminium and gold are often used and, as we have seen in chapter 4, their properties are very different.

Turning now to the rather rare instances where the output signal from the detector is derived largely from back-scattered electrons, we have seen (§4.31) that the total back-scattering coefficient increases with atomic number and varies by a large factor over the range of elements that may be present in the specimen. Here, then is an effect which may lead to strong contrast in such cases.

5.1.5. Factors in class (c)

As we have seen, the total current leaving the specimen contains both high-energy back-scattered electrons and low-energy secondaries and it is important to assess the relative numbers in these two groups. Detailed information on this matter has been given in chapter 4 and we are here concerned only with broad generalizations.

In the scanning electron microscope, the energy of the incident electrons will usually lie in the range 2–30 keV. Over this range, the secondary-emission coefficient falls with increasing primary energy, while any change in the back-scattered coefficient is much less marked. Thus, as a general rule, we may say that the ratio of secondary to back-scattered electrons will be greater at low than at high values of incident energy. For materials containing elements of low atomic number, the ratio will generally be greater than unity over the whole range of energy and may be much greater. With elements of high atomic number, the ratio may fall somewhat below unity at the highest values of incident energy.

When we turn to the circumstances under which electrons leaving the specimen are collected by the detector, we find that the ratio of secondary to back-scattered components which contribute to the output signal, can be modified in a number of ways. In the first place, potential differences can be maintained between detector and specimen and between either of these and surrounding electrodes or the wall of the specimen chamber. With applied voltages of a few hundred volts, it is possible to establish electric fields which exert a major effect on the trajectories of the slow secondary electrons. It is common practice to maintain the detector at a few hundred volts positive with respect to the specimen and the number of secondaries collected is then likely to be considerably greater than it would have been without the applied voltage. How large the gain will be depends on the size and position of the detector and on the applied voltage. It depends also on the local topographical structure of the surface. Slow electrons emitted from an area in a hollow may be largely shielded from the effect of the electric field and many of them may re-enter the surface before they can be accelerated towards the detector. For such areas the gain resulting from the application of the field is likely to be much less than for areas standing above the mean surface. Thus we may expect to observe considerable contrast from this effect, with raised areas showing bright and hollows dark. We note that contrast of this kind is similar to that which obtains when an object is observed visually.

So far we have dealt only with electric fields which accelerate slow electrons towards the detector. Such fields can, of course, be reversed to make it impossible for secondaries to reach the detector and thus to ensure that the output signal is derived entirely from back-scattered electrons. In general this is disadvantageous; it decreases the signal/noise ratio and, as a rule, it causes a reduction of contrast.

We might also consider the application of sufficiently strong fields to have a pronounced effect on the trajectories of the fast back-scattered electrons. This, however, would need voltages of the same order as that used to accelerate the incident probe and would be inconvenient, besides offering no particular advantage. Usually, therefore, the back-scattered electrons which enter the detector are those which are moving towards the detector when they leave the specimen surface. This fact must contribute to contrast between two areas of the specimen which are inclined at an angle to each other, because the number of electrons leaving the surface, per unit solid angle, directed towards the detector, is unlikely to be the same in the two cases (§4.3.1). As a rule there is no corresponding effect with secondary electrons, because the directions in which these electrons leave the specimen are rendered relatively unimportant by the application of an electric field between specimen and detector.

We have previously drawn attention to the way in which the size and position of the detector may influence some of the factors that we have been discussing. If it is desired to reduce the ratio of back-scattered to secondary electrons, the axis of the detector should not lie in the plane containing the incident probe and the normal to the mean surface of the specimen, since back-scattered electrons will usually be more numerous in this plane (§4.3.4). Again, the ratio is likely to be reduced by increasing the angle between the axis of the collector and the normal to the mean surface of the specimen. This will reduce the collection of back-scattered electrons in accordance with the Lambert cosine law (§4.3.1); it will probably have less effect on the collection of secondaries, because of the potential difference normally applied between specimen and detector.

Special effects can sometimes be produced by using a small detector or by artificially reducing the aperture of a large one, though either procedure is generally disadvantageous since it reduces signal/noise ratio. At the opposite end of the scale, when an output signal is derived from the flow of specimen current through a high resistor, all electrons leaving the specimen contribute to the signal. In such a case the contrast must certainly be different from what it would have been if the signal had been derived from the more usual collection of only a fraction of these electrons, but little experimental work on this matter has been published. Finally, it must be remembered that back-scattered electrons which do not enter the detector will strike the surface of the final lens or the wall of the specimen chamber, where they will generate secondaries. Some of these secondaries will enter the detector and, in some circumstances, they may account for as much as thirty per cent of the detector output. It is not easy to assess their effect on contrast.

From what has been said in the preceding sections, it will be apparent that the formation of contrast is a most complex business, resulting from the interplay of a number of largely independent factors. The microscopist should be aware of these factors, since he has under his control several ways in which he can influence them. He can vary the energy of the incident probe and the angle at which it strikes the mean surface of the specimen; the position, size and inclination of the detector; the application of electric fields in the vicinity of the specimen and the nature of any conducting coating that the specimen may need. He cannot be given general rules for the use of these controls, but he is likely to employ them more effectively if he has a thorough understanding of the effects which each is likely to produce.

5.1.6. Contrast resulting from electric or magnetic fields at, or near, the specimen surface

In early experiments with the scanning electron microscope it was found that the output signal from the detector would change quite markedly when the potential of the specimen with respect to its surroundings was varied by as little as 1 V. Following this observation, it was shown by Oatley and Everhart (1957) that the effect could be used to delineate areas on a specimen surface at different potentials, by producing strong contrast between these areas in the image formed by the microscope. The matter was further investigated (Everhart, Wells and Oatley, 1959) and it was found that the trajectories of slow secondary electrons, leaving the specimen and travelling towards a positively biased detector, were quite sensitive to changes in potential between the specimen and its surroundings. Thus the number of these electrons entering the detector depended on this potential and the observed contrast in the image could be explained on this basis. It was also found that the magnitude of the effect depended quite critically on the position of the detector and on the size and shape of its entrance aperture. Under favourable conditions potential difference as low as 0.1 V, between different areas of the specimen, could be detected.

The above effect has proved to be of considerable importance in the investigation of potential variations in semiconductor devices and integrated circuits, and several investigators have introduced techniques with the object of making quantitative measurements of these variations. We shall not discuss this work here, since it is of somewhat specialized application, but two related matters of more general interest will be mentioned briefly, of which the first concerns the causes of 'potential contrast'. In the early investigations it was tacitly assumed that the number of secondary electrons reaching the detector from any given area of the specimen, was determined by the relative potentials of that area, the detector and the walls of the specimen chamber. More recent work by Plows (1969) has shown that fields set up by differences of potential in the specimen itself may also be important; electrons from regions of low potential may be drawn across to more positive regions, before they can be accelerated towards the detector by the externally applied field. Any such redistribution of electrons is obviously important if quantitative measurements of potential variation on the specimen are to be made. Plows shows that the difficulty may be overcome by placing in front of the specimen surface a grid at a positive potential, which draws electrons away from the surface before redistribution can occur.

A second general technique, introduced by Oatley (1969), relates to a method of separating contrast produced by external bias in a semiconducting device, from contrast produced in other ways. A chopped incident electron probe is used, with the frequency of chopping sufficiently high for several cycles to occur while the probe is moving from one picture point to the next. The pulses of current arriving at the detector are amplified and the output is passed to two parallel electronic gates, A and B, operated in such a way that alternate pulses travel through separate channels. External bias is applied to the specimen by a square-wave voltage, whose frequency is half that of the chopping frequency and whose phase is adjusted so that signals in channel A come from a biased specimen, while for those in channel B the specimen is unbiased. The signal in each channel is passed to a detector, d.c. restoration having previously been applied. Finally, the outputs from the two channels are passed to a differential amplifier and the resulting difference signal is led to the display unit. A schematic diagram of the apparatus is shown in fig. 5.1 (a), while fig. 5.1 (b) shows the waveforms of the various voltages and currents: (iii) and (iv) are the operating voltages applied to open gates A and B respectively. With this arrangement, fairly complete cancellation of all contrast, other than that resulting from the external bias, can be achieved.

Since electrons can be deflected by magnetic fields, it is to be expected that contrast will result when such fields are present in the specimen. They might be caused, for example, by domain structure or by the non-uniform magnetization in a magnetic tape. Contrast

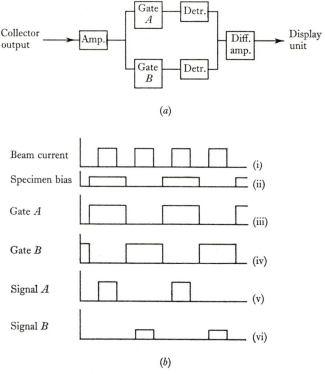

(a)

(b)

Fig. 5.1. Schematic arrangement for separating potential contrast from other forms of contrast. Redrawn by permission from Oatley (1969).

of this kind has been observed by Banbury and Nixon (1967) and more recent work has been reported by Joy and Jakubovics (1968, 1969). Banbury and Nixon (1969) have also shown that contrast, whether resulting from the topographical structure of the surface or from inherent electric or magnetic fields in the specimen, can often be enhanced by the application of additional fields between specimen and detector.

5.1.7. Gamma correction

In the preceding discussion of contrast it has been assumed that the output signal from the detector, after any necessary linear amplification, is passed directly to the cathode-ray tube of the display unit. However, the brightness of a cathode-ray tube is not proportional to the signal voltage applied to its grid and, even if it were, there is no reason to expect that the complex processes which determine the variations in signal voltage, would lead to satisfactory variations of brightness in the image. This problem is well known in television engineering, where non-linearities in light/electricity conversion at the camera tube and electricity/light conversion at the cathode-ray tube are compensated by non-linear amplification of the electrical signal in the receiver. In any non-linear process of this kind it is usually a sufficiently good approximation to write

$$\text{output} = k(\text{input})^\gamma, \tag{5.2}$$

where k and γ are constants. We then characterize the non-linearity by its gamma value and the use of a correcting amplifier is commonly referred to as gamma correction.

In the case of a television chain, the gamma values of the camera tube at the input, and of the cathode-ray tube at the output, can be measured and an amplifier to give any necessary correction to the overall gamma of the system can then be designed. In the scanning electron microscope the matter is less simple. The gamma of the cathode-ray tube used for visual display will probably be about 2.5, but little information is available about the effective gamma that should be assigned to the process by which the input signal to the detector is generated. We are not here concerned with the relation between the input probe current and the current reaching the detector for any given picture point, but with whether the variation of the latter from one picture point to another bears the correct relation to the variation in brightness that we should expect to find if we could view the specimen visually. We are interpreting visually a signal that has been produced electrically and experience is our only guide in the matter of gamma correction. When we turn to photographic reproduction, the problem is further complicated by the fact that the gamma of the photographic process depends on the film used and the conditions under which it is developed and printed. For all these reasons, it is desirable to provide, in the amplifier chain of the microscope, circuits by means of which the gamma can be varied manually over a range of, say, 1.0 to 0.25. Suitable circuits can be designed in a variety of ways by making use of the non-linearities in the characteristics of thermionic valves and semiconducting devices.

5.1.8. Other forms of signal processing

In addition to gamma correction, which is of very general application, brief mention will now be made of certain other types of signal processing which are advantageous for special purposes. Hitherto we have been principally concerned to produce, on the display tube, a 'realistic visual image' of the specimen, but this need not necessarily be our object. Sometimes we are more interested in detecting or locating certain features which are known to be present in the specimen, without regard to the general appearance of the surface. Where such features involve abrupt changes in signal current, it may be advantageous to differentiate the signal before passing it to the display unit. Or, again, a compromise may be achieved by passing to the display unit a composite voltage, obtained by adding fractions of the original signal and the differentiated version. The value of such techniques can only be assessed by trial.

Another type of signal processing consists in passing the output signal from the detector to a computer which has been programmed to extract certain information. For example, it might count the number of particles lying within a particular range of size. So far, there is relatively little experience of techniques of this sort, but they seem likely to become of increasing importance.

5.2. Stereoscopic operation

The pronounced three-dimensional appearance of scanning electron micrographs usually makes their interpretation a relatively straightforward matter. Occasionally, however, ambiguities may arise or, alternatively, more precise information may be needed about the heights of 'hills' or the depths of 'valleys' on the specimen surface. In such cases it may be useful to take stereoscopic pairs of micrographs, to be examined through a viewer in the usual way.

When considering how best to obtain a stereoscopic pair of micrographs, it must be borne in mind that the view of the specimen which each micrograph presents is that seen when looking along the incident electron probe. Moreover, the three-dimensional effect obtained, when the pair is observed through a stereoscopic viewer, is an optical illusion which depends on the interpretation

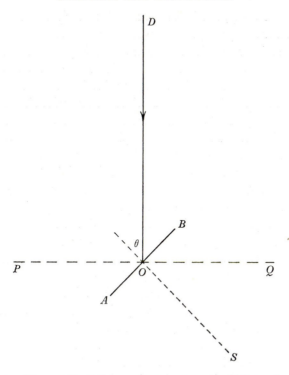

Fig. 5.2. Diagram illustrating specimen movements which may be used to produce stereo pairs.

placed by the brain on the information conveyed to it by the two eyes of the observer. In making this interpretation the brain relies on its previous experience of the appearances of three-dimensional objects. Thus, a perfect stereoscopic pair of micrographs should record the two views which the separate eyes of an observer would see, if he were looking at an enlarged model of the specimen. The micrographs will be recorded consecutively, the specimen being moved slightly between exposures. We now consider what movement of the specimen is appropriate.

In fig. 5.2, the mean plane AB of the specimen is at right angles to the paper and its normal makes angle θ with the incident probe DO, which lies in the plane of the paper. One common method of making stereopairs is to vary the angle θ by two to fifteen degrees between successive exposures. Satisfactory results are often obtained in this way, particularly when θ is nearly equal to ninety degrees. For many purposes, however, θ may be as small as thirty degrees, to facilitate collection of the secondary electrons, and the two views do not then correspond with those which an observer would commonly choose for visual examination of a surface whose normal was inclined at angle θ to his mean line of sight. In such a case he would prefer to orientate the surface so that its normal and his mean line of sight lay in a vertical plane, whereas the two micrographs produced by a small change in θ correspond to a situation in which the surface normal and the mean line of sight lie in a horizontal plane. When θ is nearly a right angle, this difficulty largely disappears but, in other cases, the optical illusion produced when the micrographs are examined in a stereo viewer may be less than perfect. When the specimen is extremely rough, a mean surface plane does not exist and stereo pairs obtained by small variations of θ are likely to be more satisfactory in this case.

The arguments just advanced suggest that a better way of making a stereo pair would be to rotate the specimen through a small angle about the axis PQ in fig. 5.2, between successive exposures. However, the specimen stage may not make provision for such a rotation. So far as is known, little work has been done to determine whether better results are, in fact, obtained by this method.

An alternative method, which has often been used, is to rotate the specimen through a small angle about the axis OS, a movement for which the specimen stage usually makes provision. This can be regarded as the resultant of rotations about DO and PQ respectively, of which the latter is the one that is wanted. At first sight the rotation about DO can be ignored, since it does not change the aspect of the specimen as seen along the incident beam. However, it does change the position of the specimen with respect to the detector and this may alter the appearance of the image in some way. As a result, it is sometimes found difficult to secure satisfactory registration of the two micrographs of a pair, when they are placed in a viewer. The effect is not usually serious if the rotation about OS does not exceed ten degrees.

Stereo pairs are particularly valuable when quantitative information is to be extracted from the micrographs. The way in which this may be done has been considered by Wells (1960) and by Lane (1969).

5.3. Resolution

5.3.1. Introduction

We define the resolution of a scanning electron microscope to be the smallest distance between two separate features of the specimen which permits these features to be reliably distinguished in the image. Such a definition may lack precision, but it has practical significance and draws attention to the fact that the resolution achieved in any given instance depends on the specimen as well as on the microscope. In particular, the contrast between the two features to be resolved must enter into the problem.

When the highest resolution is sought, a micrograph with an appreciable background of noise is likely to be obtained. In this case the noise spikes themselves may give rise to areas of light and shade in the micrograph which may be mistaken for features present in the specimen. It is therefore good practice, when any doubt arises, to take a second micrograph to confirm the existence of the features; artefacts produced by noise spikes are unlikely to occur in identical positions in the second micrograph.

5.3.2. The probe diameter

Suppose the electron probe to be scanning across a part of the specimen in which two 'black' areas are separated by a parallel-sided 'white' area of width x. If x is greater than the diameter d of the probe the output signal will rise from zero in the black area to full strength in the white area. On the other hand, if x is less than d, the change in signal strength will be reduced, but it will not be zero unless x is negligible in comparison with d. In principle, therefore, it does not seem impossible that a microscope should resolve distances smaller than the probe diameter. In practice, the above simple argument is complicated by the facts that the signal is superimposed on a background of noise and that the current density in the probe spot decreases gradually with distance from the centre. On both counts the change in signal strength, as the white area is crossed, is reduced, and experience suggests that distances smaller than the probe diameter are unlikely to be resolved.

Factors limiting the diameter of the probe have been discussed in chapter 2 and the equations there derived lead to the curves of fig. 5.3 when the typical values listed in the figure are assigned to the various parameters.

These values are not necessarily the best that can be achieved. For example, the cathode current density and temperature are those typical of a tungsten hairpin, but at least a ten-fold increase in current density and a reduction in temperature from about 2800 °K to about 2000 °K could be obtained by using a lanthanum hexaboride cathode. Then, again, by setting $B/\Delta B$ equal to ten, we have made an arbitrary choice and it is of interest to know whether the choice is a reasonable one. Here we may be guided by subjective

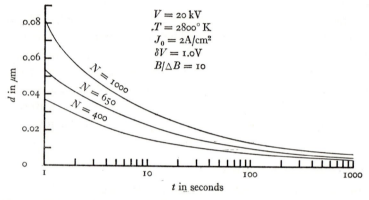

Fig. 5.3. Relations between probe diameter, number of lines and exposure time. After Oatley *et al.* In *Adv. in elec. and electron phys.* ed. Marton, **21**, 181. © Academic Press 1956.

assessments that have been made of television images, where the quality of the image is judged in relation to the ratio of the peak white signal to the r.m.s. noise. In our case this ratio is represented by $I/\sqrt{(\overline{\Delta I^2})}$ (§ 3.4.3). It is found that a ratio of about 180 is needed for the high quality picture required in a studio monitor: a ratio of thirty is adequate for a home receiver and a ratio of ten still has appreciable entertainment value. We have shown that putting $B/\Delta B$ equal to ten is equivalent to setting $I/\sqrt{(\overline{\Delta I^2})}$ equal to fifty, a ratio that would be more than adequate for the image on a television receiver. Clearly, if we are seeking the highest possible resolution, we shall be prepared to tolerate a lower ratio of signal to noise, in order to reduce the diameter of the probe.

Taking all these factors into account it appears that, under the most favourable conditions, considerable improvements on the results represented by the curves of fig. 5.3 should be attainable. A probe diameter of 0.005 μm or less should be usable without excessive length of recording time.

5.3.3. Penetration of electrons into the specimen

To obtain high resolution, an accelerating voltage of 20–30 kV is commonly used, and the incident electrons will then penetrate the specimen to depths of the order of 1.0 μm. We have seen (§ 4.5.3) that secondary electrons are unlikely to escape through the surface if they are generated at depths much greater than 0.01 μm. Thus, if the incident probe strikes the surface normally, most of the escaping secondaries which have been generated by the ingoing primaries, will leave the surface within about 0.005 μm from the point of entry of the probe. On the other hand, secondaries generated by back-scattered primaries (§ 4.5.2) may leave the surface at much greater distances from the incident probe. This effect will be still more marked when the incident probe strikes the surface obliquely: both the ingoing primaries and the back-scattered electrons may then produce secondaries which leave the surface at relatively great distances from the point of entry of the incident probe.

Another situation arises when an insulating specimen is given a thin coating of evaporated metal, to prevent charging; both the nature of the metal and the thickness of the coating are then under the control of the operator. This case has been considered by Everhart (1970), who concludes that, for biological specimens, gold is one of the best coating materials and that the coating thickness should be 0.008–0.01 μm.

The fact that some of the secondaries may leave the surface at distances up to perhaps 1.0 μm from the point of impact of the primaries is of no importance so long as they reach the detector. It is clearly possible, however, that the number of secondaries reaching the detector may be influenced by features of the surface remote from the point of entry of the probe. When this happens, the interpretation of the micrograph will be hindered and the effective resolution may be worse than it would otherwise have been.

In addition to the secondary electrons considered above, the detector input current will contain back-scattered primaries which may leave the specimen surface at considerable distances from the point of entry. These electrons contribute very little useful information about the specimen, but they do add to the background noise and so reduce the signal/noise ratio.

From what has been said it may be concluded that the penetration of primary electrons into the specimen will have an adverse effect on resolution and that the magnitude of the effect will almost certainly vary greatly from one specimen to another.

5.3.4. Experimental results

Careful measurements of resolution have been made by Broers (1969*b*, 1970), using a scanning microscope specially designed to give the best possible results. In particular, great attention was paid to the elimination of vibration and electrical interference. The minimum probe diameter was estimated to be 0.0025 μm.

Broers emphasizes the need to avoid contamination of the specimen, resulting from decomposition of pump-oil vapour by the electron stream, since this can obscure fine detail on the specimen surface. He overcame this difficulty by using ion pumps. With carefully selected specimens he obtained a resolution of 0.005 $\pm$ 0.001 μm. For a wider range of specimens a value of about 0.015 μm would probably be more representative.

BIBLIOGRAPHY

Italic numbers in square brackets indicate the pages on which the reference is cited.

Adams, J. and Manley, B. W. (1967). *Philips Tech. Rev.* **28**, 156 [*114*].
Ahmed, H. and Broers, A. N. (1972). Private communication [*58*].
Albert, M. J., Atta, M. A. and Gabor, D. (1967). *Brit. J. App. Phys.* **18**, 627 [*60*].
Amos, S. W. and Birkinshaw, D. C. (1962). *Television Engineering*, vol. 4 (London: Iliffe) [*105*].
Archard, G. D. (1953). *J. Sci. Inst.* **30**, 352 [*77*].
Archard, G. D. (1956). *Brit. J. App. Phys.* **7**, 330 [*62*].
Archard, G. D. (1958). *Rev. Sci. Inst.* **29**, 1049 [*69*].
Banbury, J. R. and Nixon, W. C. (1967). *J. Sci. Inst.* **44**, 889 [*180*].
Banbury, J. R. and Nixon, W. C. (1969). *J. Phys.* E2, 1055 [*180*].
Barnes, R. L. and Openshaw, I. K. (1968). *J. Phys.* E1, 628 [*85*].
Bassett, R. and Mulvey, T. (1969). *Zeit. f. ang. Phys.* **27**, 142 [*87*].
Baxter, A. S. (1949). *Ph.D. Dissertation*, Cambridge [*9*].
Beck, A. H. (1959). *Proc. I.E.E.* **106**B, 372 [*60*].
Beck, A. H., Maloney, C. E. and Mead, P. F. (1970). Private communication [*60*].
Becker, A. (1929). *Ann. Phys. Lpz.* **2**, 249 [*162*].
Bishop, H. E. (1966). *Ph.D. Dissertation*, Cambridge [*149, 150*].
Bloomer, R. N. (1957a). *Proc. I.E.E.* **104**B, 153 [*44*].
Bloomer, R. N. (1957b). *Brit. J. App. Phys.* **8**, 85 [*44*].
Boersch, H. and Born, G. (1960). In *4th Inter. Conf. on Elec. Microscopy*, ed. Möllenstedt, G., Niehrs, H. and Ruska, E., p. 35 (Berlin: Springer) [*40*].
Bothe, W. (1921). *Z. Phys.* **5**, 63 [*136*].
Bothe, W. (1933). *Hand. d. Phys.* vol. 22, ch. 2, p. 1 (Berlin: Springer) [*136*].
Brachet, C. (1946). *Bull. Assoc. Tech. Maritime Aeron.* **45**, 369 [*7*].
Broers, A. N. (1969a). *J. Phys.* E2, 273 [*59*].
Broers, A. N. (1969b). *Rev. Sci. Inst.* **40**, 1040 [*188*].
Broers, A. N. (1970). In *Proc. 3rd Scan. Elec. Micros. Symp.* ed. Johari, M., p. 1 (Chicago: I.I.T. Res. Inst) [*188*].
Bruining, H. (1954). *Physics and Applications of Secondary Emission* (London: Pergamon) [*157, 162*].
Butler, J. W. (1966). In *6th Inter. Congr. Elec. Microscopy*, ed. Uyeda, R., p. 191 (Tokyo: Maruzen) [*55*].
Cosslett, V. E. and Thomas, R. N. (1964a). *Brit. J. App. Phys.* **15**, 235 [*134, 135*].
Cosslett, V. E. and Thomas, R. N. (1964b). *Brit. J. App. Phys.* **15**, 883 [*134, 137, 139*].
Cosslett, V. E. and Thomas, R. N. (1964c). *Brit. J. App. Phys.* **15**, 1283 [*134, 143, 144, 147*].
Cosslett, V. E. and Thomas, R. N. (1965). *Brit. J. App. Phys.* **16**, 779 [*134, 149, 152, 156*].

Crewe, A. V. (1971). *Phil. Trans. Roy. Soc.* B **261**, 61 [*12*].

Crewe, A. V., Eggenberger, D. N., Wall, J. and Welter, L. M. (1968). *Rev. Sci. Inst.* **39**, 576 [*55, 87*].

Crewe, A. V., Isaacson, M. and Johnson, D. (1970). *Rev. Sci. Inst.* **41**, 20 [*118, 119, 126*].

Davoine, F. (1957). *Dissertation*, University of Lyons [*7*].

Dearnley, G. and Northrop, D. C. (1963). *Semiconductor counters for nuclear radiations* chs. 6 and 7, p. 124 (London: Spon) [*118*].

Ditchburn, R. W. (1963). *Light* ch. 6, p. 200 (London: Blackie) [*21*].

Dolder, K. and Klemperer, O. (1957). *J. Elec. and Control*, **3**, 439 [*40, 43*].

Dreschler, M., Cosslett, V. E. and Nixon, W. C. (1958). In *4th Inter. Conf. on Elec. Microscopy*, ed. Bargmann, W., Möllenstedt, G., Niehrs, H., Peters, D., Ruska, E. and Wolpers, C. (Berlin: Springer) [*52*].

Duncumb, P. (1969). *J. Phys.* E**2**, 553 [*86, 128*].

Dyke, W. P. and Dolan, W. W. (1956). *Adv. in elec. and electron phys.* ed. Marton, L., **8**, 89 (New York: Academic Press) [*48, 49*].

Dyke, W. P., Charbonnier, F. M., Strayer, R. W., Floyd, R. L., Barbour, J. P. and Trolan, J. K. (1960). *J. App. Phys.* **31**, 790 [*50, 57*].

Ennos, A. E. (1953). *Brit. J. App. Phys.* **4**, 101 [*130*].

Ennos, A. E. (1954). *Brit. J. App. Phys.* **5**, 27 [*130*].

Everhart, T. E. (1967). *J. App. Phys.* **38**, 4944 [*54*].

Everhart, T. E., (1970). In *3rd Stereoscan Colloquium* p. 1 (Kent Cambridge Scientific Inc.) [*187*].

Everhart, T. E. and Thornley, R. F. M. (1960). *J. Sci. Inst.* **37**, 246 [*111*].

Everhart, T. E., Wells, O. C. and Oatley, C. W. (1959). *J. Electron. Control*, **7**, 97 [*173, 178*].

Fert, C. and Durandeau, P. (1967). In *Focusing of charged Particles*, ed. Septier, A. vol. 1, chs. 2, 3, p. 309 (New York: Academic Press) [*65*].

Fontijn, L. A., Bok, A. B. and Kornet, J. G. (1969). In *5th International Congress on X-ray optics and microanalysis*, p. 261 (Berlin: Springer) [*86*].

Gomer, R. (1961). *Field emission and field ionization* ch. 1, p. 1 (Cambridge: Harvard Univ. Press) [*49*].

Gray, F. (1939). *Bell Sys. Tech. J.* **18**, 1 [*80*].

Hachenberg, O. and Brauer, W. (1959). *Adv. in elec. and electron phys.* ed. Marton, L., **11**, 413. (New York: Academic Press). [*157*]

Hadley, C. P. (1953). *J. App. Phys.* **24**, 49 [*32*].

Haine, M. E., (1957). *J. Brit I.R.E.* **17**, 211 [*39*].

Haine, M. E. and Einstein, P. A. (1952). *Brit. J. App. Phys.* **3**, 40 [*40*].

Haine, M. E., Einstein, P. A. and Borcherds P. H. (1958). *Brit. J. App. Phys.* **9**, 482 [*40, 44*].

Hanszen, K.-J. and Lauer, R. (1967*a*). *Zeit. f. Naturforschung*, **22**a, 238 [*57*].

Hanszen, K.-J. and Lauer, R. (1967*b*). In *Focusing of charged particles*, ed. Septier, A., vol. 1. ch. 2.2, p. 251 (New York: Academic Press) [*62*].

Herring, C. and Nichols, M. H. (1949). *Rev. Mod. Phys.* **21**, 224 [*31*].

Hibi, T. (1956). *Jap. J. Electronmicroscopy*, **4**, 10 [*57*].

Hollway, D. L. (1962). *J. Brit. I.R.E.* **24**, 209 [*38*].

Hughes, K. A., Sulway, D. V., Wayte, R. C. and Thornton, P. R. (1967). *J. App. Phys.* **38**, 4922 [*116*].

Hutson, A. R. (1955). *Phys. Rev.* **98**, 889 [*31*].

Joy, D. C. and Jakubovics, J. P. (1968). *Phil. Mag.* **17**, 61 [*180*].

Joy, D. C. and Jakubovics, J. P. (1969). *J. Phys.* D2, 1367 [*180*].

Kanter, H. (1957). *Ann. Phys. Lpz.* **20**, 144 [*149, 153*].

Kanter, H. (1964). *Brit. J. App. Phys.* **15**, 555 [*149, 152*].

Knoll, M. (1935). *Z. Tech. Physik*, **16**, 467 [*2*].

Knoll, M. and Theile, R. (1939). *Z. Physik*, **113**, 260 [*2*].

Kollarh, R. (1956). In *Hand. d. Phys.* ed. Flugge, S., vol. 21, p. 232 (Berlin: Springer) [*157, 159*].

Kulenkampff, H. and Ruttiger, K. (1954). *Z. Phys.* **137**, 426 [*149*].

Kulenkampff, H. and Ruttiger, K. (1958). *Z. Phys.* **152**, 249 [*149*].

Kulenkampff, H. and Spyra, W. (1954). *Z. Phys.* **137**, 416 [*149*].

Lafferty, J. M. (1951). *J. App. Phys.* **22**, 299 [*58*].

Lane, G. S. (1969). *J. Phys.* E2, 565 [*184*].

Langmuir, D. B. (1937). *Proc. I.R.E.* **25**, 977 [*14, 31*].

Liebmann, G. (1955*a*). *Proc. Phys. Soc.* **68**B, 679 [*65, 97*].

Liebmann, G. (1955*b*). *Proc. Phys. Soc.* **68**B, 682 [*65, 82*].

Liebmann, G. (1955*c*). *Proc. Phys. Soc.* **68**B, 737 [*65, 68*].

Liebmann, G. and Grad, E. M. (1951). *Proc. Phys. Soc.* **64**B, 956 [*65, 67*].

McMullan, D. (1953). *Proc. I.E.E.* B100, 245 [*8, 113*].

Martin, E. E., Trolan, J. K. and Dyke, W. P. (1960). *J. App. Phys.* **31**, 782 [*50*].

Maruse, S. and Sakaki, Y. (1958). *Optik*, **15**, 485 [*57*].

Moss, H. (1961). *J. Electron and Control*, **11**, 289 [*32*].

Moss, H. (1968). *Adv. Elec. and Electron Phys.* ed. Marton, L., Supp. 3 (New York: Academic Press) [*31, 37, 39, 43, 88*].

Mulvey, T. (1952). *Proc. Phys. Soc.* **66**B, 441 [*64*].

Mulvey, T. (1959). *J. Sci. Inst.* **36**, 350 [*78*].

Munro, E. (1971). In *Proc. 25th Mtg. Inst. Phys. E.M.A.G.* ed. Nixon, W. (London: Inst. Phys.) [*84*].

Nottingham, W. B. (1939). *Phys. Rev.* **49**, 78 [*31*].

Oatley, C. W. and Everhart, T. E. (1957). *J. Electron*, **2**, 568 [*178*].

Oatley, C. W. (1969). *J. Phys.* E2, 742 [*106, 179*].

Palluel, P. (1947). *C.R. Paris*, **224**, 1492 and 1551 [*149*].

Pelcowitch, I. and Saalberg van Zelst, J. J. (1952). *Rev. Sci. Inst.* **23**, 73 [*123*].

Picard, R. G. (1954). In *3rd Inter. Conf. Elec. Microscopy*, ed. Ross, R., p. 151 (London: Royal Microscopical Soc) [*79*].

Pierce, J. R. (1949). *Theory and design of electron beams* ch. 8, p. 116 (New York: van Nostrand) [*31*].

Plows, G. S. (1969). *Ph.D. Dissertation*, Cambridge [*179*].

Rezanowich, A. (1968). In *Symp. on scanning Elec. Microscopy*, ed. Johari, O., p. 15 (Chicago: IIT Res. Inst) [*11*].

Rose, A. (1948). *Adv. in Electronics* **1**, 131, ed. Marton, L. (New York: Academic Press) [*16*].

Sakaki, Y. and Möllenstedt, G. (1956). *Optik*, **13**, 193 [*57*].

Schwartz, J. W. (1957). *R.C.A. Rev.* **18**, 1 [*38*].

Seiler, H. (1967). *Z. Angw. Physik.* **22**, 249 [*160*].

Shockley, W. and Pierce, J. R. (1938). *Proc. I.R.E.* **26**, 321 [*110*].

Simpson, J. A. and Kuyatt, C. E. (1963). *Rev. Sci. Inst.* **34**, 265 [*37*].

Smith, G. F. (1955). *Phys. Rev.* **100**, 1115 [*31*].

Smith, K. C. A. (1956). *Ph.D. Dissertation*, Cambridge [*11*].

Smith, K. C. A. (1959). *Pulp and Paper Magazine, Canada*, **60**, T366 [*11*].

Smith, K. C. A. and Oatley, C. W. (1955). *Brit. J. App. Phys.* **6**, 391 [*114*].

Sokolovskaia, I. L. (1956). *Soviet Phys.-Tech. Phys.* p. 1147 (English translation published by Amer. Inst. Phys.) [*55*].

Speidel, R. (1965). *Optik*, **23**, 126 [*57*].

Sternglass, E. J. (1954). *Phys. Rev.* **15**, 345 [*149*].

Sturrock, P. A. (1951). *Phil. Trans. Roy. Soc.* A **243**, 387 [*77*].

Swift, D. W. and Nixon, W. C. (1962). *Brit. J. App. Phys.* **13**, 288 [*57*].

Thompson, B. J. and Headrick, L. B. (1940). *Proc. I.R.E.* **28**, 318 [*38*].

Thornton, P. R. (1968). *Scanning Electron Microscopy* (London: Chapman and Hall) [*128, 129*].

van der Ziel, A. (1952). *Adv. in Electronics*. ed. Marton, L., **4**, 110 (New York: Academic Press) [*109*].

van Nie, A. G. (1968). *Electronic Eng.* **40**, 520 [*123*].

von Ardenne, M. (1938). *Z. Physik.* **109**, 553 [*3*].

Weinryb, E. and Philibert, J. (1964). *C.R. Paris*, **258**, 4535 [*149*].

Wells, O. C. (1960). *Brit. J. App. Phys.* **11**, 199 [*184*].

Wentzel, G. (1927). *Ann. Phys. Lpz.* **69**, 335 [*135*].

Whiddington, R. (1912). *Proc. Roy. Soc.* A, **86** 365 [*143*].

Wiley, W. C. and Hendee, C. F. (1962). *I.E.E.E. Trans. Nucl. Sci.* NS-9, 103 [*114*].

Woroncow, A. (1947). *Proc. I.E.E.* 93pt.3A, 1564 [*91*].

Worster, J. (1969). *Int. J. Electronics.* **27**, 49 [*54*].

Zworykin, V. K., Hillier, J. and Snyder, R. L. (1942). *ASTM Bull* **117**, 15 [*4*].

INDEX